Project Development in
the Solar Industry

Project Development in the Solar Industry

Edited by:

Albie Fong & Jesse Tippett

with contributions from:
Arturo Alvarez
Jeffery Atkin
William DuFour III
Perry Fontana
William Hugron
Jason Keller
Tyler M. Kropf
Michael Mendelsohn
Brett Prior
Scott Reynolds
Pilar Rodriguez-Ibáñez
Katherine Ryzhaya Poster
Alfonso Tovar

Text edited by:
Jane Tippett & Santiago Quintana

CRC Press
Taylor & Francis Group
Boca Raton London New York

CRC Press is an imprint of the
Taylor & Francis Group, an **informa** business

A BALKEMA BOOK

Cover Art by: Michael T. Kachanis – www.michaeltkachanis.com

Published by:
CRC Press/Balkema
P.O. Box 447, 2300 AK Leiden, The Netherlands
e-mail: Pub.NL@taylorandfrancis.com
www.crcpress.com – www.taylorandfrancis.com

© 2013 Taylor & Francis Group, London, UK
CRC Press is an imprint of Taylor & Francis Group, an Informa business

No claim to original U.S. Government works

ISBN 13: 978-0-367-57661-5 (pbk)
ISBN 13: 978-0-415-62108-3 (hbk)

Visit the Taylor & Francis Web site at
http://www.taylorandfrancis.com

and the CRC Press Web site at
http://www.crcpress.com

Typeset by V Publishing Solutions Pvt Ltd., Chennai, India

Library of Congress Cataloging-in-Publication Data

To Mom & Dad, Jason, Karen, and Grandpa

Albie Fong

To all my teachers, especially Mrs. Oexner, my mom and Dad, my grandparents, and my wife and Kids

Jesse Tippett

Contents

Acknowledgments

We would like to especially thank our industry colleagues, and co-authors, for their contributions to this publication as everybody provided vital insight to making this reality. Perry, Tyler, William, Jason, Arturo, Katherine, Alfonso, Jeff, Bill, Michael, Scott, Pilar, and Brett, we hope to continue increasing the presence of the solar industry with you. We would like to thank Jon Albisu and Emmanuel of Albiasa Solar for believing in us by allowing us a chance to get into the solar industry; without that experience, we would never have become experts in solar project development. We would like to thank Jane and Santiago for providing technical editing and providing your honest feedback. We are truly grateful and thankful to Janjaap and CRC Press for their vision, understanding, and professionalism in working with us over these years to come up with a comprehensive book on solar project development. Finally, this publication would not have been possible without the support of Claudia and Karen in allowing us to spend countless nights and hours committed to compiling this book.

List of figures

List of tables

This book and electronic enhancements

THIS BOOK

Currently the Solar Energy Sector is one of the world's fastest growing markets. As such, this burgeoning market deserves dissection and dissertation. To date, most solar related texts have focused on installation of rooftop solar systems and projects on a smaller scale. We have not yet seen a comprehensive book that outlines the heavy-weight topics that we will investigate in the following chapters. *Project Development in the Solar Industry* will prove to be the premiere resource for companies and individuals wishing to become involved, work, and succeed on a large scale in the solar industry.

It has been written with these objectives in mind to date that we, Albie Fong and Jesse Tippett set out to collaborate with a team of international experts to provide comprehensive information not just about technical details but also about the basics of the business aspects of solar development and everything in between. Whether you are in a related field, a project developer, a policy maker, an engineer, or just curious about the large scale solar industry, this book defines the process and background to one of the most important, fastest and most exciting dynamic markets of our time.

We hope you enjoy this book and get involved with renewable energy; it is good for our economy, our environment and perhaps even your pocketbook.

ELECTRONIC ENHANCEMENTS

In the following text we make reference to the book's companion website www.solarbookteam.com. This website provides contact information for all authors to this text and access to the key resources highlighted in this book. This tailored media platform provides supplemental and exclusive information that is up-to-date with the present state of the solar industry. We invite anyone interested in posing new questions or opportunities to contact us through this service.

About the authors*

ALBIE FONG

Albie is currently a Senior Manager of Project Execution with Talesun Solar where he works to deploy Talesun manufactured photovoltaic modules into commercial and utility scale projects. Technical evaluation of solar equipment, EPC management, and creation of strategic partnerships are other aspects of his role.

Before his time with Talesun, Albie started the North American operations of Albiasa Solar in late 2008 and has held positions of both Chief Project Engineer and Managing Director of the company. During his tenure at Albiasa, Albie has been involved directly in many levels of the solar value chain for both utility scale PV & CSP projects: engineering, energy modeling, project permitting, manufacturing supply chain management, product marketing, and financing.

Prior to his solar career, Albie worked in San Diego with OptiSwitch Technology Corporation, a group focused on the engineering of high power semiconductor switches with applications in the defense and biofuel industries. A native of the California Bay Area, Albie obtained his B.S. in electrical engineering from the University of California, San Diego with a concentration in financial investment. Traveling, snorkeling, and baseball have been activities Albie enjoys when not otherwise engaged saving the world with green energy.

JESSE TIPPETT

Jesse is a business development professional currently with Aries Power & Industrial where he works with their international team to develop renewable energy and provide engineering design and construction services to projects. Prior to joining Aries Jesse focused on technology bankability, project acquisitions and the development and energy marketing aspects of renewable energy projects with GCL Solar Energy and Albiasa Corporation.

Jesse became involved in renewable energy during his undergraduate studies at Worcester Polytechnic Institute, where he graduated with a B.S. in mechanical engineering and concentrations in Aerospace and Spanish. While studying he completed an award winning renewable energy project in Namibia, Africa with the R3E that implemented recyclable materials into low cost housing to decrease energy usage. Jesse has an MBA from the University of New Haven. Aside from contributing to the solar energy field, Jesse has worked in the aerospace industry with UTC's Sikorsky Aircraft and completed a research project at NASA's Glenn Research center. When not working, Jesse enjoys reading non-fiction, family time with his lovely wife and two beautiful children, adventure, traveling, surfing, and any combination of the aforementioned.

ARTURO ALVAREZ

Arturo Alvarez currently serves as a project manager for Sisener Engineering in Phoenix, Arizona. In his current position Arturo is primarily responsible for the coordination of engineering documentation and main point of contact for ongoing projects that range from construction to advanced levels of development. Previously Arturo served as Albiasa Corporation's Project Development Manager. In this role, Arturo was primarily responsible for pre-development efforts and preliminary engineering for Albiasa's various CSP and PV projects. Additionally Arturo led all North American, Chinese, and Middle Eastern suppliers sourcing and supply chain management for the Albiasa parabolic trough for utility scale systems. Furthermore Arturo was a liaison between Albiasa's Spain based engineers and the local regulations and procedures.

Arturo's key involvement in the renewable industry include innovative solutions for the adaptation of alternative fuel and the viability of biogas projects for disadvantaged communities in South America in 2007. This successful project addressed the lack of fossil fuels and took advantage of the available resources while educating the public about the biogas. Other involvements include the de optimization of distributed energy systems using solar PV.

Arturo graduated from Santa Clara University with a Mechanical Engineering degree. He was an active member of the American Society of Mechanical Engineers and involved in ASME design projects.

JEFFERY ATKIN

Jeffery R. Atkin is a partner with Foley & Lardner LLP where he is chair of the Solar Energy Team and a member of the Energy Industry Team. His areas of practice cover a broad range of business and financial matters, including renewable energy, project finance, private placements, mergers and acquisitions, joint ventures, real estate development and equipment procurement and leasing.

Jeff's experience in renewable energy and project finance includes representing developers, investors, lenders and landowners in the construction, development, acquisition and financing of renewable energy generation facilities including: wind, solar, hydro, geothermal and biomass facilities. He is a frequent speaker and author on all types of renewable energy development.

Jeff received his J.D., magna cum laude from Brigham Young University, where he was a member of the Brigham Young University Law Review. He received his B.A., summa cum laude, from Southern Utah University.

When not working, Jeff enjoys just about any event that will be getting him outside in the sun – from golfing to surfing to digging weeds.

WILLIAM DUFOUR III

William D. (Bill) DuFour III is an associate with Foley & Lardner LLP and a member of the firm's Transactional & Securities Practice and the Energy Industry Team. His areas of practice cover a broad range of business and financial matters, including renewable energy, project finance, mergers and acquisitions (domestic and foreign) and general corporate matters. In particular, Bill has extensive experience in wind and solar project development, as well as acquisitions and financings of such projects.

Bill earned his law degree from The George Washington University Law School (J.D., cum laude), where he was a Thurgood Marshall Scholar. He holds a master's degree in education from Loyola Marymount University (M.A., summa cum laude) and a bachelor's degree in economics and political science from the University of Southern California (B.A., cum laude).

PERRY FONTANA

Perry H. Fontana, QEP has been permitting and developing energy facilities for 35 years. As Founder, President and CEO of Fontana Energy Associates he assists clients in the licensing of both renewable and conventional generation technologies. Perry has witnessed the evolution of the permitting process both domestically and internationally and he has a solid understanding of how the environmental review process can have a fundamental impact on the viability of a project.

Perry was formerly Vice President-Projects for Ausra, Inc. where he was responsible for site acquisition and development. At Ausra he acquired sites and completed early stage development for over 2 GW of solar thermal generation. Prior to his work at Ausra he was Vice President of URS Corporation, responsible for their Electric Power Business in the western U.S. and international operations. At URS he directed the site selection and permitting for natural gas, coal, geothermal, wind, solar and gasification projects totaling over 10 GW of installed generation. He has worked in 25 countries and has served as an expert for several independent power producers, the US Department of Energy, and numerous multinational lending institutions.

When not looking for that perfect and easy-to-license project site, Perry enjoys time with his family and their two English Mastiffs.

WILLIAM HUGRON

William Hugron has been in commercial real estate for 36 years, 10 years in property management, 12 years as a Vice President in asset management for Citicorp, Cal Fed, CAL REIT, and Peregrine Real Estate Trust and the remainder of his career in commercial brokerage and consulting. He was engaged by the bankruptcy courts in implement strategies to take Common Wealth Equity Partners out of bankruptcy and turn the entire portfolio around. It eventually was renamed the Peregrine Real Estate Trust.

He has completed multi land transactions consisting of infill land and land assemblage. William has completed land assemblage assignments consisting of over 6,000 acres for Victoria Homes in Victorville, Lewis Operating Company in Adelanto, and Strata Land Investors in Apple Valley. He has also represented buyer & sellers and tenant & landlords in the Victor Valley and in Southern California. Throughout his career he has been able to reposition and dispose of special assets especially while working for syndicators and financial institutions throughout the Western United States. He was also given an award one of the largest office leases in 2008 by SIOR and CoStar for a 10 year, 65,000 square foot office lease valued at $24,800,000. He is consistently one of the top producers at Ashwill Associates and was a top producer when at the Charles Dunn Company. Besides practicing commercial real estate brokerage he also engages on commercial real estate consulting.

He has contributed to many trade magazine articles and has lectured and given classes on commercial real estate.

JASON KELLER

Jason has been working in commercial real estate since 2001. From 2002–2004, Jason focused on office, industrial, and retail leasing, and developed his negotiating skills while representing both Landlords and Tenants. Since 2004, Jason has been representing residential developers, solar developers & private investors in the acquisition and disposition of land, raw and entitled, located throughout Southern California.

Jason's true expertise lies in sourcing land. Using state of the art technology, Jason will exceed client expectations, delivering the right sites at the right price.

In his free time, Jason enjoys exercising and spending time with his beautiful wife and 5 children. He and his family live in Corona, California.

TYLER M. KROPF

Tyler M. Kropf is a commercial real estate broker at Ashwill Associates, Inc. where he directs land acquisition and disposition for lending institutions, public companies, private investment groups and individual owners.

Early in his career his primary work was representing residential developers and merchant builders. His background in large-scale developments gave him experience to fill a more recent advisory position with utility scale solar development companies. He has been included in several professional publications to speak on the subject of land and renewable energy development.

Tyler graduated from Brigham Young University with a bachelor's degree in Facilities Management and a Minor in Business Management. Tyler is fluent in Spanish and enjoys playing tennis and living in Newport Beach, California with his wife and two daughters.

MICHAEL MENDELSOHN

Mike Mendelsohn has over 20 years of experience in various facets of the energy industry. Mike is currently a Senior Financial Analyst with the National Renewable Energy Laboratory, and specializes in financial structures and sources of capital for renewable energy projects. He also leads the development of evaluation tools to assess project cost of energy, assesses the impact of financial structure on utility-scale solar projects, and is currently evaluating the use of securitization to improve access to capital. His analyses and other products can be found at: www.financeRE.nrel.gov

Prior to joining NREL, Mike worked as a senior associate with the non-profit Western Resource Advocates, a senior consultant with Levitan and Associates, and an economist with the Massachusetts Department of Public Utilities. Mike holds a M.S. in Energy Management and Policy from the University of Pennsylvania and a B.A. in Computer Science from Ithaca College.

BRETT PRIOR

Brett Prior is a Senior Analyst with PHOTON Consulting. Previously, he covered the solar industry at GTM Research, the market research arm of Greentech Media. Brett has published reports on Concentrating Solar Thermal Power (CSP), Concentrating Photovoltaics (CPV), Polysilicon, and Project Finance for Renewable Energy.

An expert in utility-scale solar, he has been quoted in numerous publications including: The New York Times, Reuters, Forbes, and Bloomberg. He has also contributed articles for GE's Ecomagination website. Brett earned an MBA with distinction from London Business School. Before graduate school, Brett served as a senior equity analyst at Bishop Street Capital Management, a mutual fund company in Hawaii. Brett holds a BA in Political Science and International Studies from Yale University, and he is a Chartered Financial Analyst (CFA).

SCOTT REYNOLDS

Scott Reynolds is a key member of Beecher Carlson's Energy Practice. His primary responsibilities involve designing, marketing, and implementing innovative insurance programs for Energy clients, especially in the renewable and alternative energy space. In addition to managing projects and responding to the day-to-day service needs of clients, he has expertise in crafting creative solutions to business challenges.

Prior to joining Beecher Carlson, Scott was a Senior Vice President at Marsh where he led Marsh's Alternative Energy Practice for the West Zone, U.S. Solar and Geothermal Practices and Offshore Insurance team for Chevron Corporation. Before that, Scott served as the Director of Risk Management for FPL Group, Inc., a diversified holding company that owned Florida Power & Light Company and had interests in commercial banking, real estate, cable TV and cogeneration venture. Scott has extensive experience in the field of utility risk management, both through his work at Florida Power and Light and as chairman of numerous utility industry committees. Before joining FPL Group, Inc., Scott worked for Earl V. Maynard & Company Insurance Broker, specializing in electric and gas utilities and private energy firms.

PILAR RODRIGUEZ-IBÁÑEZ

Pilar Rodriguez-Ibáñez holds a Master's Degree in Environmental Policy from the University of Illinois at Urbana-Champaign. She is also a PhD candidate in Sustainable Development from Complutense University of Madrid, Spain. She is a former local representative in her home State of Veracruz in Mexico. Her research interests have focused on Solid Waste Policy, Environmental Taxation, and Energy Policy.

Currently, she is a Graduate student in Public Policy at ITESM's Escuela de Graduados en Administración Pública y Política Pública (EGAP) in Monterrey, Mexico. When not working Pilar enjoys spending time with her wonderful husband and two lovely boys.

KATHERINE RYZHAYA POSTER

Katherine Ryzhaya Poster is a Vice President in the Structured Transactions group at Evolution Markets. Based in the firm's San Francisco office, Ms. Ryzhaya Poster works with wholesale market participants on all matters relating to power marketing, energy sales, contract negotiations, and policy development. Ms. Ryzhaya Poster joined Evolution from Pacific Gas and Electric Company (PG&E) where she served as Principal in the Energy Procurement Group and was responsible for leading negotiations on utility-scale renewable energy contracts. Prior to PG&E, Ms. Ryzhaya Poster worked in finance at RBC Capital Markets and at UBS. She received her MBA, with a concentration in Finance, from Duke University.

In her free time, Katherine builds toy forts, puts on puppet shows and scales the hills of San Francisco with her stroller (and baby).

ALFONSO TOVAR

Alfonso Tovar is a Solar Engineer currently working with Black & Veatch, where he specializes in the assessment of established and emerging solar technologies and systems including: photovoltaic (PV) systems design, design review of utility scale and large rooftop PV systems, modeling and performance evaluation of PV systems, training on PV technologies, cost analysis of solar systems, and solar policy. Alfonso began his career as a Solar Engineer a decade ago, designing and installing small-scale off-grid PV systems for SolarQuest, a non-profit organization. He then obtained a graduate degree at the University of California, Merced, where he conducted research on micro-concentrating solar power systems using non-imaging optics. After graduation, Alfonso worked with two PV systems integrators of large commercial and utility scale photovoltaic systems before joining Black & Veatch.

Chapter 1

Why renewable energy?

At present, a growing population of over 7 billion people inhabits the planet. The growth rate is expected to slow each year, however, a population of somewhere between 7.5 and 10.5 billion people by 2050 is anticipated, according to the Population Division of the Department of Economic and Social Affairs of the United Nations Secretariat (2009). Natural gas, coal, and petroleum power plants are responsible for over 65% of all electricity today. Many accept these conventional sources of energy, also known as fossil fuels, as reliable long-term energy sources.

As the world's population increases so does the world's energy usage. If fossil fuels remain the dominant form of energy generation, the world as we know it will likely continue on its present course of increased consumption. The question is "In what form is that course, and is it a sustainable one?" Al Gore's now famous presentation, *An Inconvenient Truth,* highlights historical data showing not only a global warming trend, but also its link to the greenhouse gas carbon dioxide, CO_2, a known byproduct of fossil fuel use. While determining the causes of CO_2 creation is an issue that is being evaluated, this will not be the only environmental problem the world will face if fossil fuels remain our dominant source of energy generation.

Tenth of a degree annual average temperature changes due to global warming have been both relatively subtle up to today and challenged by both sides of the global warming debate. Hence, building an argument for renewable energy based solely on "solving" climatic change or global warming is hard to quantify and highly controversial. But what if there were highly observable and almost indisputable figures available that are directly related to use of conventional energy? With fluctuations not just in the tenths of a percent but in the ten or even twenty percent rate of change, these figures exist and we have to look no further than the volatile historical prices of our conventional energy commodities. Take the retail price of gasoline or the market price or natural gas used in electricity generating power plants. Increases of over 10–20% in one month have been seen in gasoline prices and natural gas has been no less dynamic with both up and down swings of 100% in 1–2 year periods (Arndt, 1989). While gasoline and diesel prices are nearing record highs today in 2012, the price of natural gas is the lowest it has been in 10 years. Some may not see an issue with historically dynamic or volatile prices, especially when they are low. However, price volatility is seen an important factor in economics. According to a report by the Center for American Progress, energy price volatility has a very negative effect on the economy. Data presented in their May 2011 report titled *Not Again The Summer Vacation Gas Price Roller Coaster on the Move Again* states that "… energy price

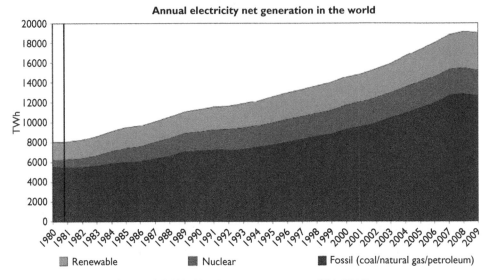

Figure 1.1 Worldwide energy generation (EIA, 2011).

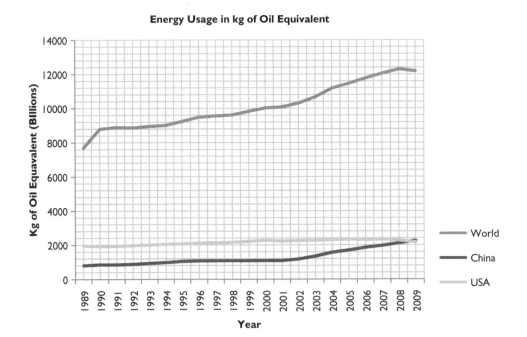

Figure 1.2 Energy usage in oil equivalent (World Bank Database, May 2012).

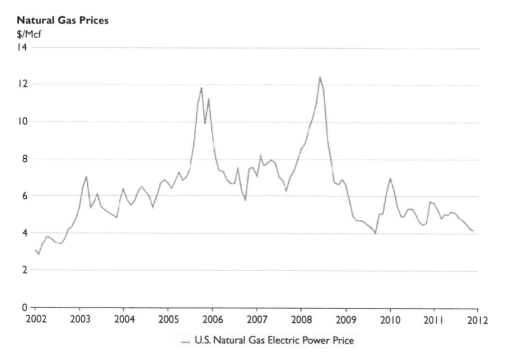

Natural Gas Prices

$/Mcf

Figure 1.3 Historic natural gas prices (EIA, June 2012).

volatility – wide fluctuations in gasoline and other prices – lead families and businesses to delay investments." Furthermore, "…the economy suffers because of less spending and investment." Solving the problem of energy price volatility will have benefits for the economy and our environment.

While scientists debate the predicted outcomes of humans and their fossil electricity generation on our climate, it remains certain to all that fossil fuels prices are anything but predictable. In contrast, renewable energy is much more certain and is perhaps without the price volatility. In support of this statement let us briefly investigate the business model of a natural gas power plant compared with a solar energy power plant of similar size. Both cost money to construct and operate and gain revenue through the sale of electricity. Natural gas power plants must purchase fuel to generate electricity for the lifespan of the plant, while a solar energy power plant, once built, will make energy from sunshine alone. The natural gas power plant must hedge against future fuel prices to reduce the risk to its cash flow, conversely the solar power plant has great certainty built into its revenue structure as decades of historical weather data are statistically analyzed to arrive at conservative assumptions for the amount of solar irradiance available. Electricity from natural gas is procured on a spot market in some power plants, and as a consequence during periods of high fuel cost, a natural gas power plant may be forced to operate at

Weekly Retail Gasoline and Diesel Prices

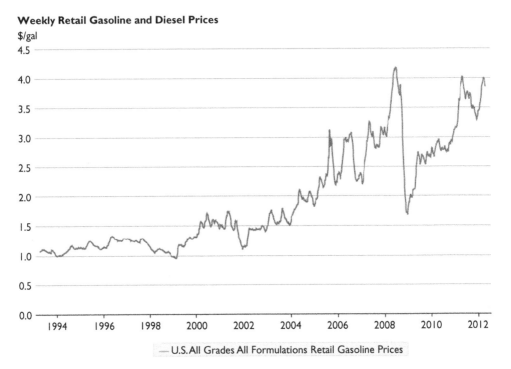

Figure 1.4 Historic retail gasoline and diesel prices (EIA, June 2012).

a loss if they are locked into selling energy at a price that is too low to make a profit due to their variable energy costs. This situation is not only theoretical but it has happened to Calpine, a natural gas power plant developer with gigawatts of power plants installed who declared bankruptcy in 2005 due to this exact situation (Douglass, 2006).

Considering the broad spectrum of 21st century technology, we do not profess that renewable energy generation, or solar alone, should make up 100% of the world's energy mix. However, the benefit and value of the avoided cost of price uncertainty and arguably lower pollution profile of renewable energy is not yet reflected in today's global energy mix. Major factors in price volatility are supply and demand and as our population continues to grow, demand increases for typical energy sources. If nations wish to avoid the economic and possible climatic effects the status quo may conjure; they must embrace renewable energy if only for its intrinsic lack of price volatility. Finally, while the cost of many types of renewable energy are still higher than the entrenched generation models we have used for the last 100 years, we suggest that when the value offered by renewable energies is truly considered, the population will all see change for the better. These benefits include less price uncertainty, lower total cost when considering all impacts, higher sustainability and hopefully a healthier planet.

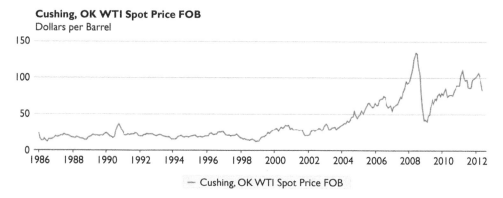

Figure 1.5 West Texas Intermediate (WTI) spot oil prices from 1986 to 2012 (Thompson Reuters, 2012).

Figure 1.6 Historical coal prices (EIA, June 2012).

1.1 "WHY NOT" SOLAR ENERGY

While there are several reasons why solar energy can benefit our economy and our planet, there are also counterpoints to the implementation of solar energy. At present, solar energy is not the lowest cost method to generate electricity and for that reason alone, it has not seen widespread adoption. The price of solar generating equipment and amount of labor required to install that equipment correlates to the high production cost of solar energy. The cheaper you can buy the equipment and any reductions to the cost of labor means that you can afford to sell electricity at a lower price.

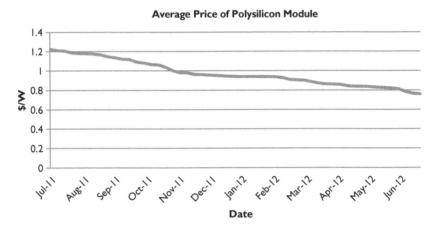

Figure 1.7 Historical polysilicon photovoltaic module cost breakdown showing a 40% price decrease over 12 months (*Source*: data from PVinsights.com, July 2012).

Prices of photovoltaic panels, also known as PV panels, have decreased significantly just over the last few years. PV panels represent the largest percentage of a project's cost and their decreased cost correlates to the ability to offer electricity to utilities at a more competitive price. PV panel price reduction results from several key factors: increased supply to the market due to increased manufacturing capacity, alternative technology options that are less dependent on silicon material in PV products (i.e. new thin film and concentrating photovoltaic), increased efficiencies PV panels and improved automated manufacturing methods.

A widespread goal of the solar energy industry is to produce electricity at grid parity. Grid parity is defined as a level where alternative energy is producing electricity at costs equal to that of power purchased from the grid. For the solar industry to meet grid parity, it is the responsibility of individual solar companies to reconcile creating a profitable business with continued cost reductions while maintaining competitive value against other forms of energy generation. Solar technology companies have the best leverage to come up with new ideas to reduce costs which may come in many forms. These methods have included using less expensive materials or more efficient use of material, higher efficiency generation technology, making a product that is easier to assemble or easier to install in the field. With the solar industry continually making these strides to reduce costs and increase performance and value, it maintains high interest from all stakeholders that can benefit from the reduction in cost to produce solar energy.

Solar energy facilities, when not installed on unused space like rooftops, requires large amounts of land. When these types of facilities have been proposed in sensitive environmental areas such as protected wildlife habitats, their large 'footprint' has been met with resistance from various groups. The desert southwest of the United States consists of vast tracts of arid land sitting unoccupied, without residential, agriculture, or planned urban development. Much of this vacant desert land is situated in an area of such powerful solar resource which has led to significant development of large solar projects, especially in southern California. Depending on the technology, the amount of acres allocated for every Megawatt (MW) of solar energy generally ranges from 5–10 acres per MW. For instance the SunPower 250 MW project, California Valley Solar Ranch, in San Luis Obispo County is planned to cover 1,966 acres of land (California Valley Solar Ranch, 2009). Unless the technology changes drastically there is not likely to be a significant decrease in the land required. However, prudent siting of large projects is the best way to satisfy stakeholders.

Nuclear, natural gas, and coal-generated electricity all have the benefit over solar power generation which is that they are not intermittent resources. Photovoltaic and concentrated solar power energy are intermittent resources that only generate energy while the sun shines or clouds are not passing overhead.

Intermittent solar intervals have been addressed using state of the art meteorological predictions and the coalescence of the solar generation curve with typical peak energy demand and usage. Daily peak demand periods in the United States, over the course of a year, typically occur between late afternoon and early evening when people arrive home from work and start turning on all their appliances (an interesting contrast shows a second demand peak electricity usage of electricity in the late evenings in Spain). Assuming the electric vehicle industry matures as expected, the near future will see a new surge of electricity demand for electric vehicle users plugging in their vehicle when arriving home from work. A full charge can take anywhere from 6–16 hours, depending on the voltage rating of the charging equipment and the size of the vehicle battery. When utility companies see a significant usage of energy from its customers, they must have power generating resources to compensate for this demand in electricity.

If solar power plants do not have the expected solar resource available to generate because of weather intermittence, the utility company would be left without power that it may have expected to receive from a solar power plant. Weather intermittence can be largely mitigated by spacing solar power plants across a large area to minimize any local weather anomalies and by maintaining a healthy mix of generating technologies.

Aside from the large footprint of even a single solar facility, one of the largest problems with the development of solar power plants is the effect of its implementation on the environmental characteristics of the surrounding area. Utility scale solar projects require a significant amount of land area for the same net power compared to conventional fuel plants. The effect of utility scale solar power plant implementation on threatened or endangered plant and animal species and cultural sites is of high concern to the environmental community. Furthermore, there are high concerns of the visual impact of the eventual operating solar power plant, air quality during construction, migration pattern disruption, wildlife endangerment due to environmental disturbance, pre-historic and historic cultural resources impact, and many other topics that are studied in an environmental impact statement of projects.

While California has recently seen many new solar power plant installations, the state having one of the best solar resources in the United States, it also a state with a long list of protected species. Some threatened and endangered species have come into the public spotlight due to the conflict of installing solar power plants in habitat areas of those species. Some of these species include the San Joaquin kit fox, desert tortoise, Mohave ground squirrel, blunt nosed leopard lizard, and Swainson's hawk. How solar projects are able to co-exist with these species is highly project dependent. The mitigation procedures (if any) are addressed during the permitting process with the County, the State, and any public lead permitting agencies (the group in charge of a permitting process when several agencies are involved). In many cases, solar projects cannot co-exist with potentially endangered species and appropriate approvals will not be received for a solar project; this leads to a flawed project and has to be abandoned unless there is an alternative design that can be implemented. In some cases where there is potential conflict with existing species, mitigation procedures can be implemented to allow projects to move forward while being within the reasonable conservation boundaries established by the environmental community. Sometimes the proposed mitigation efforts can pose severe modifications to a project that destroy its technical or economical ability to continue with developmental efforts. Therefore, extensive discussions should occur between the project developer and the permitting agencies to ensure there are solutions that balance the goals of both sides.

A prime example of the ongoing effort of utility scale solar development and significant interaction with the environmental community is one of the largest upcoming Concentrating Solar Power (CSP) projects to be built: the 392 MW Bright-Source Ivanpah project. The developer and equipment supplier, BrightSource Energy, was required to buy mitigation land for the desert tortoise, requested to monitor each desert tortoise found for five years while testing blood samples for respiratory diseases, and was said to have hired up to 100 biologists to work on site, as of June 2011, to monitor the species (Wesoff, 2011). These efforts to satisfy environmental mitigation procedures were all in addition to scaling down the size of the project to receive full approval and authority to commence construction by the California Public Utility Commission (CPUC). Through the permitting process through the CPUC, BrightSource agreed to reduce the plant gross power output from 440 MW to 392 MW (Prior, 2010). This action led to a smaller project footprint further minimizing environmental impact to the desert tortoise population. Regardless of the mitigation efforts, environmental species will be affected directly or indirectly and that cannot be avoided. Threatened and endangered species will die during the construction of these facilities and it may not be known for many years what the true effect of large solar power plants will have on the various desert species. This co-existence between solar power plants and environmental species will continue to be a highly debated topic as long as solar projects require hundreds or thousands of acres to achieve viable economies of scale.

Chapter 2

Technology basics

The sources of renewable energy depending on interpretation are somewhat wide ranging since they consist of energy derived from numerous natural resources. Generally speaking, energy generated without using fossil fuels (coal, natural gas, or petroleum) or that is non-carbon emitting is said to be green, renewable energy. This broad definition could allow for conventional nuclear reactors to be classified as renewable energy. However, typically legislation excludes nuclear as renewable energy due to obvious controversy. Some also argue that biomass, along with waste to energy processes, should not be classified as renewable due to carbon being released into the atmosphere during combustion. In the United States, the federal government has defined incentive eligible renewable energy technologies as the following:

- Solar Water Heat
- Solar Space Heat
- Solar Thermal Electric
- Solar Thermal Process Heat
- Photovoltaics
- Wind
- Biomass
- Geothermal Electric
- Fuel Cells
- Geothermal Heat Pumps
- Combined Heat and power (CHP)/Cogeneration
- Solar Hybrid Lighting
- Fuel Cells using Renewable Fuels
- Microturbines
- Geothermal Direct-Use

2.1 SOLAR PHOTOVOLTAICS

2.1.1 The photoelectric effect

Solar photovoltaics, or PV, would not be able to work were it not for the photoelectric effect. The photoelectric effect is a physical process whereby the energy of photons of various wavelengths strikes a material and the solid material gives off energy in

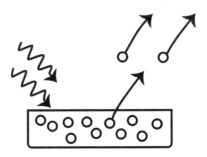

Figure 2.1 When light, also known as two photons, hit a crystal lattice in a solar cell, it causes electrons to be "knocked loose".

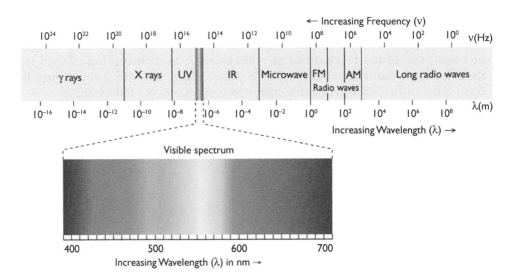

Figure 2.2 Wavelength of light.

the form of electrons. Nearly all materials exhibit this effect for certain wavelengths of electromagnetic radiation. Photovoltaic solar technology absorbs photons in the range of the electromagnetic spectrum we call light; light that consists of electromagnetic radiation with wavelengths of 10^{-8} to 10^{-3} meters and a frequency of 10^{16} to 10^{11} Hz.

In solar PV technology, the photoelectric effect is responsible for converting light into electricity. This occurs first at the solar panel, also known as a PV module, where direct current (DC) is generated. A typical panel produces between 200–300 Watts (W) of power at a standard test condition of 1000 Watts per meter squared of direct solar irradiation depending on the type and quality of the panel.

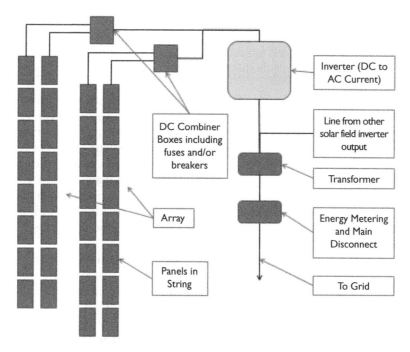

Figure 2.3 General PV block diagram layout.

PV panels are usually connected together in series of ten to twenty modules consisting of groupings called strings. The number of modules in a string is limited by the open circuit voltage, which is specified by the module manufacturer. Strings are then connected together at the DC junction box forming an array. Arrays are then grouped together through a high current cable that ultimately terminates at an inverter. The inverter serves to convert the direct current to alternating current; a form of electric power that businesses and residences can utilize. Voltage entering the inverter is normally at a maximum of 600 volts for commercial and residential systems; voltages get up to 1000 volts for European and US utility scale systems. The inverter outputs polyphase (1, 2 or 3 phase) alternating current, usually between 250 V_{ac} to 380 V_{ac}, for utility scale systems. A utility scale system transforms the electricity that the inverter outputs to the desired voltage (typically 12–120 kilovolts). A residential or commercial inverter may have integral transformers to transform electricity directly to usable voltage. Energy meters are usually placed at the outputs of both the inverter and primary transformer. Energy totals measured in these energy meters are used to calculate energy payments to a system owner where applicable.

Today the majority of photovoltaic energy generating systems installed utilize polycrystalline or monocrystalline type solar panels; however several other PV technologies exist that utilize a similar architecture as previously discussed. One of these types is thin film solar panels that have potentially better performance in low light and high heat than polysilicon panels. Concentrating PV (CPV) utilizes optics (mirrors or lenses) to concentrate light onto a small multi-junction diode or polysilicon cell, hence utilizing less expensive material in an effort to reduce costs. Additionally, several other technologies are currently being developed, and one worth mentioning is the luminescent solar concentrator, basically sheets of plastic that would allow light in but not out. Similar to CVP, luminescent solar concentrators have a relatively small amount of active material inside these inexpensive plastic sheets which serves to convert light to electricity.

The main purported benefits of solar PV in comparison to other renewable technologies are that it is reliable and predictable from both energy production and system reliability standpoints. Additionally, while the systems do require maintenance, a general lack or low number of moving parts means that maintenance can be minimal.

Solar PV facilities require the same types of approvals and permits to be constructed as other energy generating projects; however, the main difference from other types of energy developments is the land requirement. Typically solar PV facilities require five to seven acres of land for every 1 megawatt (MW) of peak power. The required amount of land needed per megawatt of net capacity will decrease as the efficiency of solar panels increases.

Solar energy systems have an energy profile that coincides with the sun's daily cycle. Energy begins to be generated just before dawn, peaks at noon, and stops just after sunset. On an annual basis, the majority of a system's energy is generated closest to the local summer solstice, while very little cumulative energy is generated near the local winter solstice or typical winter months. While individual clouds are unpredictable, energy generated by solar PV Systems can be predicted with a high degree of accuracy even on a day before or an hour before basis.

Solar PV systems show generous improvement in efficiency gains; however, the typical sunlight into electricity conversion efficiency of ten to near twenty percent is far from high. Polysilicon solar modules conversion efficiency for instance is typically twelve to twenty percent on a per area basis. System losses in the wiring, conversion losses at the inverter, and transformation losses in the transformer serve to dissipate or lose twenty to thirty percent of the energy from the panels. Minimizing system losses is an area of study currently attracting a significant amount of research and development.

Solar PV power plants with proper maintenance are forecasted to produce energy for twenty to thirty years; however, each year solar panels become slightly less efficient (0.7%) due to oxidation. Additionally, inverter components often need to be fully replaced every ten years due to component lifecycle limitations. In 2012, solar PV systems vary in cost from as low as $2.00 per watt peak for large utility scale systems, to as high as $4.00 to $5.00 dollars per watt for high-efficiency, dual axis tracking systems and Roof top.

2.2   SOLAR THERMAL

2.2.1   The Rankine cycle

The Rankine cycle is the core phenomenon in the vast majority of power systems; ninety percent of the world's energy generation is in systems as wide ranging as nuclear, coal, natural gas, and renewables utilizing the Rankine cycle. The basic principle in the Rankine cycle

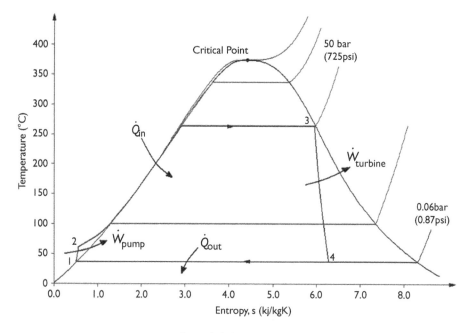

Figure 2.4 Rankine cycle.

is that incompressible fluid expands when heat is added and that this heat or energy can be extracted in the process of this expansion.

Figure 2.4 depicts the different processes of a Rankine cycle from start to finish. First water (the working fluid in this example) is pumped into a boiler (between points 1 and 2 in Figure 2.4) where heat is added (between points 2 and 3 in Figure 2.4) until steam at a high pressure and temperature is produced. This steam is then allowed to expand through a turbine (between points 3 and 4 in Figure 2.4) or other device (piston engine in the case of our car, see the Carnot cycle) where the heat energy is converted into mechanical energy in the form of rotating machinery. The steam that exits the turbine is at a lower temperature (but not so low as to allow the steam to turn to liquid water) and lower pressure than when it entered as its energy (enthalpy) has been extracted, and is then passed through a cooler or heat exchanger (between points 4 and 1 in Figure 2.4) where the steam is then condensed back into liquid water so that it can be pumped with a low amount of energy (liquids are easier to move than gases) back into the boiler (between points 1 and 2 in Figure 2.4) so the cycle can start all over again.

Fluids other than water can be used in the Rankine cycle and are chosen based on the conditions at which they vaporize or condense. Geothermal energy systems use fluids similar to Butane due to the fact that it boils at a lower temperature than water and therefore can make use of the thermal energy available from geothermal resources. Additionally, Rankine cycle systems are closed loop systems that lose very little, if any, of their working fluid.

Energy generation, in solar thermal systems, is accomplished through capturing and collecting sunlight and converting it to heat to be used in a thermal cycle like the Rankine or Stirling cycle. There are two main categories of solar thermal systems. First, the non-concentrating systems primarily found on small residential or commercial installations that are used to heat water for personal use. Second, the concentrating solar power type in which mirrors or lenses are used to concentrate the sunlight onto a smaller area to achieve a high temperature on some working fluid medium. For high temperature steam or electricity generation, concentrating systems are used. The most common of the concentrating types consists of power towers, parabolic trough systems, parabolic dishes, and Fresnel systems.

2.3   PARABOLIC TROUGH SYSTEMS

The parabolic trough may be the modern world's oldest example of the application of solar energy for utility scale power generation. American inventor Frank Shuman demonstrated the technology in a working plant in Cairo Egypt in 1913 producing 44 kW peak power. While this is not very large by modern day standards, Shuman did not use materials more exotic than cast iron tubes for heat collection and glass mirrors to reflect and concentrate the suns light. It is impressive to think that this technology has been in operation for nearly 100 years and has not departed in form from the initial design.

Today, parabolic trough systems are not necessarily the most cost effective solar thermal energy generating system, but they do have the most installed power capacity globally (as of mid-2012). Plant output in the United States alone totals over 500 MW, with 354 MW of capacity having been in operation for nearly 25 years at the Solar Energy Generating Systems (SEGS) facilities in southern California. Power Tower technology is quickly vying to be the most installed technology with firms such as Solar Reserve and BrightSource getting ready to break ground on several hundred MW of projects.

The parabolic trough power plant consists of two main parts: the solar field and the power block. In the solar field, curved mirrors are arranged to form a trough having a parabolic cross section called a solar collector assembly (SCA). These SCAs focus the sun's light onto a receiver tube called a heat collection element (HCE) situated at the parabolic trough's focal point. The tube at the center is hollow and contains a heat transfer fluid (HTF), commonly a type of oil, which absorbs the thermal energy transferred to the tube by the focusing of incident sunlight. These SCAs are grouped into sections and arranged in north to south arrangements of interconnecting troughs. These troughs track the sun's movement (commonly with hydraulic actuators) along one axis of rotation turning from east to west to track the sun's daily movement. Troughs are then arranged into a series arrangement called loops so that the HTF temperature gain can achieve the desired change of temperature after passing through the loop. The HCEs are connected in such a way that as the sun's light is focused onto the HTF, pumped through the inlet of the loop and passes through to the outlet, it increases in temperature. Groups of loops representing a Concentrating Solar Power (CSP) solar fields are positioned so that their outlet and inlet orifices can be connected to large diameter HTF header pipes. In order to keep the HTF within its operating

THE ELECTRICAL EXPERIMENTER

H. GERNSBACK EDITOR
H. W. SECOR ASSOCIATE EDITOR

Vol. III. Whole No. 35   MARCH, 1916   Number 11

The Utilization of the Sun's Energy

Years Ago Man Endeavored to Make Practical Use of the Energy Contained in the Sun's Rays—Even Tesla, the Electrical Wizard, Has Patented a Sun Motor, While the Shuman-Boy's Engine and Sun Boiler Has Developed 100 H. P. There Is Great Promise Held Forth to Future Engineers Who May Work on This Problem.

IT has been given to astrophysicists to measure the heat generated by the sun and calculate the force emanating from it. We know that the surface of our luminary gives out a heat estimated to be about 6,000° centigrade, and that its light equals that of 27,000,000,000 candlepower a quarter of a mile away. The heat which the earth receives from the sun in the course of a year would suffice to melt a belt of ice about 55 yards in thickness extending clear around the earth. Only the 2,735-millionth part of the total energy given off by the sun reaches our earth and, if this

were lacking, our planet, with all its thousandfold life, its thick forests and fruitful plains, would turn into a dead, rigid ball of rock, for the average annual temperature, which is now one of 13° centigrade of warmth for Europe, would, without the heat of the sun, sink to 73° centigrade of frost, it is calculated.

Every sort of light with which we illuminate our home when the greater light has sunk beneath the horizon, every fire that warms us when the solar rays can no longer do so, is a product originating in the sun. The chip of wood with which

the untaught son of nature brightens his hut, the twigs with which he stokes his fire, what are they but pieces of trees that grew in the sunlight? The gas of the city dweller, the coals with which he heats his house and from which the gas has been sucked, what are they but transformed sunbeams? The coal in the grate is the petrified wood of perished forests that covered the earth's surface millions of years ago, and flourished in the rays of the same sun that ripens our corn to-day. Petroleum, that mysterious earth-oil, comes from the bodies of millions of dead and

A Successful 100 H.P. Sun Power Plant Located at Meadi, on the Nile, Egypt.

Figure 2.6 Modern day parabolic troughs tracking the sun, picture taken looking north.

Figure 2.7 SEGS parabolic trough power plant in Southern California.

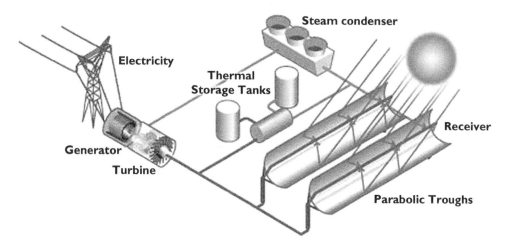

Figure 2.8 Hot and cold molten salt thermal energy storage tanks system.

maximum and minimum temperatures, the pumping of HTF is closely regulated by powerful electrically controlled pumps. Solar parabolic trough systems consume up to twenty percent of the energy that is produced to pump the HTF through a solar field potentially containing dozens of miles of piping in large facilities.

Headers coming from the solar fields (with HTF typically exiting loops at 390C, while fluid returning from the power block to the solar field is no lower than 290C) are routed back to the power block where heat exchangers transfer the thermal energy to create steam directly or into a thermal storage system to be utilized at a later time. The steam that is generated is then expanded through a turbine to spin a generator and create electric power.

One of the main benefits of solar thermal is that the energy can be easily stored for long periods of time with only a small additional cost. In the thermal energy storage portion of a parabolic trough plant, headers returning from the solar field with hot HTF are piped to a group of heat exchangers where the thermal energy in the HTF is allowed to transfer to a thermal storage medium. Typically, this medium consists of a molten salt, which is a mixture of potassium and sodium nitrate that liquifies at temperatures above 200C. Two very large tanks exist in this system and when a plant operator wishes to store energy, molten salt is pumped from the cold tank (280C) to the hot tank (380C) through the HTF to the molten salt heat exchanger. Thus, this process takes thermal energy from the HTF and stores it in the molten salt. A plant operator may wish to release or dispatch the stored energy for a various reasons: a cloud passes over the field reducing solar energy output or in order to shape the generation curve to meet daily energy delivery requirements. In exchanging its thermal energy to the HTF, the hot molten salt is simply shifted from the hot tank through the heat exchangers and back to the cold tank. In this scenario it gives thermal energy to the HTF, which makes its way to the power block.

2.4 POWER TOWERS, DISH SOLAR THERMAL SYSTEMS AND SOLAR WATER HEATING

Solar water heaters are installed all over the world on both commercial and residential locations. As energy from the sun is converted directly from light to heat, there is little loss in efficiency compared to systems that convert the sun's light to electricity through DC to AC conversion, and later voltage transformation, where losses occur at each stage. While the majority of systems generate water for comfortable bathing or home heating, it is possible to reach temperatures high enough to cook food.

Solar thermal residential and commercial systems work by pumping water from an insulated storage tank through a solar collector where the sun's light heats a matrix of tubes through which the water needing to be heated passes through. After it passes through this matrix, it returns to the storage tank where it resides until it is either discharged for use or passed back through the solar collector to be reheated.

Power tower systems work similarly to parabolic trough systems and while they do not yet have the same volume of installed capacity in operation throughout the world, they do have several intrinsic advantages causing them to quickly gain an installation advantage. The first solar power tower systems were installed in Daggett (California) at a facility called Solar One commissioned in the early 1980's.

The power tower consists of a field filled with mirrors that follow the sun on dual axis trackers (these units are called heliostats), a large centrally located tower called the receiver, a power block, and a thermal storage system. Unlike parabolic trough systems, where the sun's light is concentrated in many different places in the field, power towers focus the entirety of incident light onto this central collector tower and achieve temperatures of up to 700C. Additionally, having one receiver or collector in power tower systems greatly minimizes the amount of complex assembly required. The focus is so strong that a bird flying through its path is likely to be vaporized in midflight.

As higher temperatures are reached, compared to parabolic trough systems, the type of fluid utilized in the receiver is normally either water for direct steam generation to spin the turbine in the power block or molten salt. Systems utilizing molten salt in the receiver further save on ancillary equipment by often having thermal storage directly built into the basic system and thus saving on system costs. Similarly to parabolic trough power plans, thermal storage in power towers allows for energy to be stored in the daytime and dispatched anytime of the day or night.

While many forms of CSP have been investigated and will continue to be developed (such as linear Fresnel and hybrid CSP/CPV solutions) stirling dish collectors are theoretically the most energy efficient of the solar thermal systems due to the use of the Stirling cycle. Stirling dish collectors consist of a parabolic dish of mirrors that follow the sun on a two axis tracking system. A Stirling engine placed at the dishes focus serves to convert the thermal energy to electricity by mechanically moving a generator. The Stirling cycle is similar to the Rankine cycle discussed earlier; however, the working fluid in lieu of steam is often a gas in a closed loop (such as hydrogen or helium) and has a higher theoretical efficiency. Parabolic dish systems have not seen widespread implementation largely due to the fact that the operation

Figure 2.9 Solar thermal water heater on a residential rooftop.

Figure 2.10 Solar power tower in operation, note the highly focused light near the power towers collector.

Figure 2.11 A parabolic dish collector and engine.

and maintenance of the many thousands of individual Stirling engine towers required to generate utility scale power is extensive. The parabolic dish system may eventually see more widespread implementation should the cost of operation and maintenance issues be mitigated.

Preface: Development section

In any part of the world, solar energy projects are simply five piece puzzles:

1) The **Land** or place where the project will be installed.
2) The **Permitting** process required to obtain permission to construct and operate the project.
3) The Power Purchase Agreement (**PPA**) or Energy Off-take Agreement that defines the project's revenue stream.
4) The **Interconnection** process which allows the project to transmit energy over existing energy transmission infrastructures.
5) The **Design** of the solar field that evolves throughout the development of the project and influences how efficiently every aspect of the development is completed.

The Project Development section of this book will teach the reader not only what goes into each of these processes but also how to manage the processes effectively. Sometimes it's not making the puzzle that is the difficult part, but making sure you have all the pieces at the right time before you start putting it together.

The following is a list of this sections chapters and authors:

Project permitting by *Perry Fontana*
Renewable energy land by *William Hugron, Jason Keller and Tyler M. Kropf*
Transmission by *Arturo Alvarez, Jesse Tippett and Albie Fong*
Energy off-take and power purchase agreements by *Jesse Tippett*
Renewable energy credits by *Katherine Ryzhaya Poster*
Developer tools by *Albie Fong*
Design considerations of photovoltaic systems by *Alfonso Tovar*

To contact any author and for more information and updates on solar energy please visit: www.solarbookteam.com

Chapter 3

Development: Project permitting

Perry Fontana

3.1 THE DEVELOPMENT AND PERMITTING LANDSCAPE

As renewable generation technologies became commercially available and utility-scale projects emerged, an urban myth arose. This myth postulated that development of renewable energy projects would be easier than other types of generation. The main foundation of this myth was that there would be a "perfect storm" of support from various political advocates or elected officials, governmental regulators, environmental groups and the public. In addition, the site selection process would be easier given that there would no longer be the need to coordinate diverse types of supporting infrastructure such as locating close to fuel pipelines or railroads, water supply, or disposal systems. Clearly, there are large areas bathed in sun or blown with wind, where "all" you needed to find was transmission. As this myth grew, it even expanded to include the vision that transmission projects would be built to facilitate renewable generation assets without substantial effort by the generation owner.

Like most of you, my parents always told me some things are just too good to be true. While parts of the renewable generation development process are easier and some transmission planning initiatives show progress, the development process for renewables is fraught with challenges, many of which are all too familiar to developers. This chapter examines the development and permitting process for utility sale projects in the context of the tensions that have arisen between the drive for renewables (and the associated reduction of dependence on fossil fuel-based generation) and concerns over land-use policies, environmental impacts and the impact of intermittent renewable resources on the stability of the grid.

Before we explore the details, it is helpful to look back a bit and set the stage for where we are today in 2012. During the mid-to-late 1990s we saw profound changes in the energy generation industry. Where once generation was fossil fuel-based and permitted, licensed, constructed, and operated by utilities; deregulation and the rise of independent power producers produced a flurry of development activity, where speed of execution in development and permitting were key to enhancing project value. In fact, the development process became much more closely aligned with an ultimate goal of producing a commercially viable project. It was no longer good enough to just complete the process. Project developers now needed to optimize development and permitting at every step.

At the same time, a number of renewable technologies accelerated their transition from research to commercial deployment. Solar technologies ranging from photovoltaic to concentrating photovoltaic to solar thermal began to deploy, in a manner similar to development of many natural gas generation projects. Independent developers and power producers led development efforts with a view to signing power sale agreements with utility customers. As these projects advanced through the development process they inevitably came to the permitting process. In solar generation's drive to reduce their costs many embraced a "bigger is better" philosophy, resulting in a number of proposals for sites covering thousands of acres of vacant public land for a single project. The urban myth that solar development is easier quickly evaporated into the harsh reality that development and permitting of any generation facility is a complicated process, where multiple and often conflicting views, opinions, regulations and policies must be managed on a daily basis. This is clearly seen by following the development and permitting progress of the numerous projects in process as we go to press. It is an important, and perhaps unfortunate, fact of development that many projects fail. There is much to learn by studying these cases of failure and we have attempted to include many of these lessons in this chapter.

However, all is not chaos and many projects are successful. While many things are not clear-cut, the development landscape does offer several important guideposts for a successful project. These include considerations for site selection, identifying key environmental criteria and determining the best approach to coordinate the efforts of multiple regulatory agencies. It is important to understand that not all agencies will view a project in the same light. Where an air quality agency may see clear benefit from a solar generating facility, a wildlife agency may see the potential for serious impacts. Where some see open desert space with no residences others see a pristine environment laden with cultural history that should be preserved. Acknowledging these different viewpoints during project planning is critical.

In this chapter we will discuss these conflicts and approaches to managing them while not losing site of the end goal; a financeable project that is delivered on schedule and within budget. While we acknowledge that every project is different and may suffer some, the challenges we explore here will begin with a suggestion or "roadmap" for development, highlighting areas that impact most if not all projects in some way. We will then expand on various approaches to project permitting. Lastly, we will discuss how the various non-governmental organizations (NGO) and other stakeholders impact the development and permitting process.

3.2   THE DEVELOPMENT PROCESS – AN OVERVIEW

The term "Development" means a lot of things to a lot of people. In the context of an electrical generation facility it is important to realize that the goal of development should be to provide as much value to the project as possible. That is, the end result of development should be a project that not only can be constructed and operated efficiently but can be financed as well. A successful project should be attractive to equity partners and lenders and provide a good rate of return to the developers. In order to do this, the development process must include considerations related to engineering and design, commercial structures for off-take, interconnection to the grid, and the

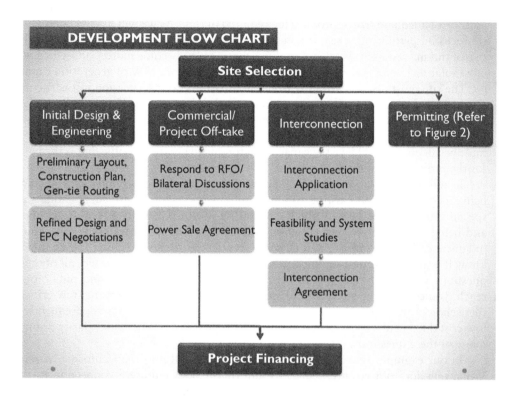

Figure 3.1 Development flow chart.

permitting and approval process. In this chapter we will provide only an overview of the complete process as further details are presented in other chapters. We will delve into the details of the permitting process here, as this is one of the most risky and dynamic aspects of development.

All development starts with a basic concept of the project. This concept may be based on perceived market opportunities, specific requests or bid solicitations by customers, or the opportunity to replace generation assets that are being retired. Regardless of the factor that leads someone to commence development, the subsequent process can be illustrated in the schematic in Figure 3.1. Clearly this "roadmap" is general but it illustrates that the development process has a lot of interrelated moving pieces, several of which may be in flux at the same time. When asked what makes a good developer I often respond that it is a combination of vision (eye on the end goal) and the ability to engage in the fine details simultaneously.

Once we have our project concept, for example a solar generation facility with a capacity of 20 megawatts (MW) or less to be located so as to serve utility customers in California, the first order of business is to start the site selection process. The site is the foundation of the project and selecting a good one will facilitate all other aspects of development. Selecting a bad one will kill the project, and sometimes that process is costly and painful. Selection of a good site considers all aspects of development.

Most importantly the site must accommodate the project concept. For solar projects it must be of adequate size to produce the required output. As we will discuss later when we focus on permitting, a major issue that confronts utility-scale solar projects is the size of the site required. While technology continues to become more "land-efficient" most solar technologies require between 5 and 10 acres per MW of generation capacity. The site must also be within reasonable proximity to transmission infrastructure possessing adequate capacity to carry the output to the target market as well as other infrastructure components such as water supply, access roads, and communications services. Interconnection to these services must be cost effective. The solar resource must be adequate to achieve the capacity factor and generation profile that is the basis for the project's economics.

In addition, a good site should minimize many potential constraints. These include being inconsistent with existing zoning requirements, being identified as habitat for special status species, being of high cultural significance to Native American tribes, and being in close proximity to sensitive receptors. These constraints involve some judgment calls. For example, how close to a residential area is too close? How can you know if a site has cultural significance? While you may never fully answer all of these questions you may have prior to actually commencing permitting and public outreach, you can greatly enhance your site selection efforts by recognizing how critical these choices are to the project's success and doing the leg work required. Frequently a low cost document survey can go a long way to providing some initial clarity around many of these questions.

In our example above, the project concept is to sell into the California market. While this does not require a site in California it does require a connection to the California Independent System Operator (CAISO), either directly or by "wheeling" the power on another system that then connects. Given the desire to be cost competitive, the best sites would offer a direct connection. This leaves us a very large initial suite of potential sites, both on private and public lands. By integrating transmission information, we can narrow our choices significantly if we focus on lands near transmission infrastructure that can accommodate the project generation with reasonable upgrade costs. These "candidate site regions" can then be further evaluated by using Geographic Information System (GIS) tools to overlay information on land use designations, habitat considerations, flood potential, topography and numerous other considerations.

You must continue to reduce the scope of potential sites until candidate regions become candidate sites by comparing the results of the GIS screening with the project concept and objectives. At this point there is no substitute to good old-fashioned legwork. Candidate sites should be ranked and the top sites should be inspected to verify the GIS information and to refine the determination of site suitability. It's important to gather as much site-specific data as possible at this stage because you may end up choosing among multiple sites that all appear feasible. Picking the best one may not be clearly obvious just from GIS studies and high-level investigations.

Project proponents should be careful to identify more than one site in case the landowner is not interested or wants too much money in exchange for site control. In the case of privately owned sites, it is not uncommon for developers to commence initial negotiations for a site option (either to lease or purchase) on multiple sites. However, the faster you can conclude the negotiations for your preferred site the better, as the time and cost to progress negotiations can be significant. Dealings with private landowners can also shed light on potential opposition from neighboring landowners. Likewise, it is not uncommon for developers to file initial applications for multiple public sites. However, the cost of advancing multiple sites for the same project is significant, and so deriving the preferred site in a timely manner should always be the goal.

While the process outlined in Figure 3.1 shows site selection proceeding before all other activities, selecting a good site (as we have discussed) includes considerations of design criteria, commercial structures, and environmental permitting issues. By making the site selection process rigorous and inclusive you can minimize delays and increased costs.

Later in this chapter we will review the permitting process in greater detail. While permitting issues tend to garner a lot of press and cause developers to lose sleep, project development has other very important pieces. Once the site has been selected and secured (or in the case of public land, an application has been accepted) site-specific engineering and design should continue. It may be helpful to think of engineering and design in three phases. First, there is conceptual engineering that supports site selection and preliminary estimates of cost. Conceptual engineering is important to ensure that the project concept can be realized at a reasonable cost and to support site selection. Preliminary engineering continues to refine the design by placing the project on the site. It is helpful early on to have a site layout, transmission routing for off-site interconnections (if required) and a preliminary determination of how much site preparation work will be needed. It is also not too early to look at site access with respect to construction equipment and equipment deliveries. Many favorable solar generation sites are located in remote areas and the cost to provide adequate access may not be trivial. Preliminary engineering should also include a construction plan, as project permitting requires estimates of equipment usage, expected schedule, and logistics such as routes to be used by construction crews and delivery trucks. Finally, refined engineering will be required to support the selection of an Engineering, Procurement, and Construction (EPC) contractor.

The EPC contract is a critical part of project finance and needs to be negotiated in close coordination with the permitting process. Large-scale solar projects typically have significant and highly detailed mitigation and monitoring requirements that apply to construction. These must be explicitly included in the EPC contract or the potential for permit violations or negative public and agency relations can have significant adverse impacts on the project. In extreme cases, there is the potential for project construction to be halted while issues of non-compliance are addressed. There are all too many examples of EPC contractors and project owners fighting over change orders resulting from a failure to adequately understand and plan for such requirements. It is important that the engineering staff be integrated into the overall development team and that communication lines be maintained. What engineering may view as an optimization of subtle design change may be significant to the project permitting and to public outreach efforts.

The commercial aspects of development are discussed in detail in other chapters of this book. However, a key aspect of defining the conceptual project and site selection process is the commercial structure of the project; primarily the planned market for the project output. It is never too early in the project process to aggressively pursue customers and secure off-take agreements. As solar project development has boomed we have noticed a significant number of projects that appear to have taken the position that if they secure a site and advance permits a customer will show up. Sadly several projects have recently failed or been shelved for lack of a power sales agreement. There has truly been a solar development boom in the southwest United States and there is evidence it is increasingly spreading geographically. Many utilities are becoming much more comfortable that they will meet their Renewable Portfolio Standard (RPS) requirements or goals. This has made them less likely to engage in bi-lateral discussions outside their more formal procurement processes. In addition, utility Request for Offer (RFO) solicitations have received overwhelming responses resulting in hundreds of bids and a dramatic fall in prices being offered. Taken together this has led to a much more competitive environment for solar developers. Without an off-take contract in place it is very unlikely that a project can be financed.

The interconnection process is currently one of the most dynamic aspects of solar project development. The boom in intermittent renewable generation including wind and solar has raised alarms at several utilities as they look at transmission system impacts and what network upgrades are required. This is understandable when you look at data points such as the interconnection queue for the CAISO. The interconnection process is also discussed in more detail in other chapters. However, it is a critical aspect of project development.

It is important to note that the interconnection process is in a state of flux in many places. In some cases there has been a separate process for small (e.g. less than 20 MW) projects. However, recent trends appear to be moving towards a single interconnection process for all projects regardless of capacity. As discussed above, the access to adequate transmission infrastructure without costly system upgrades or interconnection charges is a key part of site selection. In order to fully understand the interconnection and network upgrade aspects of the project it is necessary to make an application to the utility or system operator. In some cases such applications can only be made at certain intervals (once or twice per year). Once an application is made a typical first step is a feasibility study followed by a system impact study. Studies continue until an interconnection agreement is reached.

This process may take years to be completed until the interconnection costs and schedule are finalized. It is not uncommon for the early cost estimates and schedule (i.e. during the feasibility stage) to be very conservative. In fact these estimates may be so conservative as to skew the project economics. Clearly this impacts the project pricing and power sale negotiations. Having a qualified transmission planner on the development team is important, especially in light of the changes in the interconnection process and the queue of projects competing for space on both existing and planned transmission.

Clearly there are a lot of moving pieces, and we haven't even covered the permitting aspects yet. So how is a lead developer to manage the process? The most important thing to realize is that one person cannot do it all. A good developer builds a strong team, involves them in a cooperative decision making process and allows them appropriate responsibility so that they can have some ownership in the project. No solar generation facility development is ever going to go "according to plan." The ability to respond and adapt can mean the difference between project success and failure.

The importance of the team deserves some attention. As mentioned earlier in this chapter, a good developer keeps their eye on the overall vision (or objective) of the project. This sometimes leads to a tendency to try to do all the tasks. We also noted that development is all about the details and hopefully, the discussion above has convinced you that there are a lot of them. So how does a developer lead the process, own the vision of the project and not miss an important detail? It isn't rocket science but these simple concepts often get overlooked. First the team members should understand the vision and objectives for the project. It's up to the lead developer to communicate effectively and often. Secondly pick your team members for their expertise, which may vary by region, and value their input. They can make you look really good or really bad. Lastly, the team should expect and prepare for change. There are a lot of project management tools that can help. Lastly, recognize that development is a journey. Enjoy it.

So much for Development 101, onward to permitting.

3.3   THE EVOLUTION OF PERMITTING

The utility scale solar industry owes much of its current growth to the environmental movement. Although technology has continued to advance and costs have come down, it's hard to imagine the current market without strong policy and public support. However, as many solar project developers are learning, solar projects do not get a free pass when it comes to permitting. The southwest of the United States has numerous large projects currently in permitting and most face strong opposition from some environmental groups and regulatory agencies. Many of these concerns have been translated into litigation challenging recent permit approvals, placing project schedules (and potentially the projects themselves) at risk. Before we discuss the permitting process it is useful to have a quick history discussion as it provides useful insights into some of the drivers that motivate permitting agency staff, NGOs, and the public.

Environmental concerns regarding air and water pollution have been around for over 100 years. In the United States the framework for many landmark environmental regulations like the Clean Air Act began to come into effect in the early 1960s. However, the early versions of these regulations focused on research to try to establish health-based standards and criteria, and to reducing pollution from existing sources. This all changed in the late 1960s and early 1970s. While many cite the concern following the Santa Barbara, California oil spill in 1969, I believe that this was more of the final straw than the ultimate driver for newer and different environmental regulations. It has been my experience in working in many developing countries that environmental awareness goes hand in hand with an increased standard of living. As the pace of industrial development increased and the impacts of this development became clearer, there was a strong public desire to see stricter regulation.

While there are too many regulations to discuss without creating a whole separate book, two merit a closer look at their evolution. The first is the National Environmental Policy Act (NEPA), enacted in 1969. Title I of NEPA contains a Declaration of National Environmental Policy which requires the federal government to "use all practicable means to create and maintain conditions under which man and nature can exist in productive harmony." (USEPA, 2011) NEPA also established the Council on Environmental Quality (CEQ) to develop guidelines and procedures. The CEQ is expected to periodically issue amendments to its guidelines in an attempt to modernize NEPA. The bottom line for developers is that any major federal action, such as a land lease from the Bureau of Land Management (BLM) or project funding from the Department of Energy, must be supported by a detailed environmental review. The establishment of NEPA and similar regulations marks a shift from reactive policies to a proactive program where impacts must be analyzed and mitigated.

Not to be outdone, many states have adopted similar programs that require a similar level of analysis for projects on non-federal lands. A well-known example is the California Environmental Quality Act (CEQA), which was enacted in 1970 in direct response to NEPA. Much like the NEPA, guidelines have continued to develop and evolve. The result is that the environmental impact analysis requirements for most states differ, sometimes significantly, from the federal program. Many utility-scale solar projects have both a federal requirement and a state requirement.

The second regulation that bears a history lesson is the Endangered Species Act (ESA), established in 1973 with the stated goal of protecting imperiled species from extinction resulting from development without adequate conservation. The U.S. Fish and Wildlife Service (USFWS), and the National Oceanic and Atmospheric Administration (NOAA) are responsible for the administration of the ESA. A key program under the ESA is the "listing" of species that face a threat of extinction. Once a species is listed, the responsible agencies work to reverse the decline of the species, and ensure that development and growth do not put a species (or in some cases a local population) at jeopardy of extinction. The achievement of this goal has evolved in several ways, but a key consideration for developers is that there exists a formal process for the USFWS and/or NOAA to be consulted during a lead agencies review of the NEPA document for a project. While not discussed in detail here, a similar consultation process exists for cultural resource issues. Thus, the environmental review for any project must include detailed consideration of impacts to species covered by the ESA.

As you would expect, many states have adopted similar programs. As in the case of the overall environmental review regulations, there can be significant differences between the federal and state programs.

So what can we learn from the evolution of these and other regulations? The most important thing is that a utility-scale solar project is very likely to require comprehensive environmental review regardless of whether the site is on public or private land. This review is highly detailed and the "lead" agency will look for the project to mitigate impacts so as to comply with the intent of the regulations. Also of great importance is that there are many other agencies charged with protecting endangered species, cultural resources, air quality, water quality and others. It is not uncommon for resulting environmental documents to run well over one thousand pages. Planning for the time, effort, and cost of such a document and for the coordination among involved agencies is a key aspect of development.

3.4   PERMITTING A UTILITY-SCALE SOLAR GENERATION FACILITY

As discussed above, the permitting process is very extensive, complicated, and has a lot of moving pieces. Figure 3.2 presents a very high-level flowchart of the permitting process. Each box in the flowchart could be broken down into much more detail. For the purposes of discussion the flowchart illustrates the major pathways.

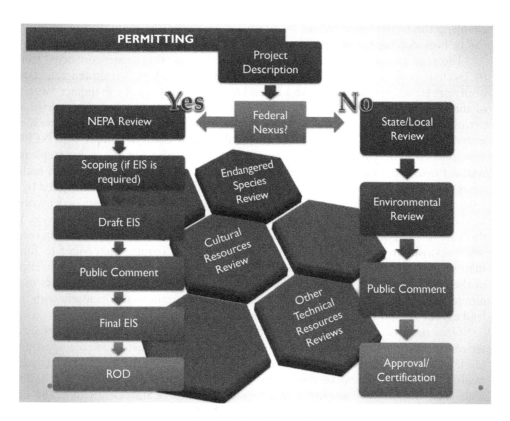

Figure 3.2 Permitting guide.

Regardless of which environmental review process you must follow (federal, state/local, or all), the first thing that should happen is to produce a clear project description. Many developers rush to have a pre-application meeting with their expected lead agency before they have such a description in place. This leads to a meeting where the agency asks a lot of questions for which answers have not yet been defined. The result is a feeling on the part of the agency that the project is illdefined and that they do not have to move forward. The project description should clearly state the project's objectives as these may limit requests to examine alternatives that are not feasible. However, the examination of feasible alternatives that may reduce impacts is an important part of environmental review so the project description should include these. Developers must take care to only include alternatives that you can execute. Serious problems can result if an agency favors an alternative that the project cannot incorporate, such as an alternative transmission line route. In addition to setting out the project location and objectives, the project description should provide conceptual design-level detail. This should include a location map, layout, type of technology (but not necessarily specific vendor), and a very high-level construction plan. This plan should discuss access routes, expected peak construction workforce estimates, and the expected schedule. It is important to not speculate on the fine details since changing these later on may lead to questions. However, the more information that can be provided the better the guidance you are likely to receive from the agencies.

The next step in permitting depends on whether your site, or any connected facility (such as your transmission interconnection, access road, etc.) is on federal, state, or private land. You may also have a federal nexus if your project will impact "waters of the United States" to the extent that you trigger a requirement for an individual permit from the U.S. Army Corps of Engineers. Funding from a federal agency, such as the Department of Energy, can also trigger the requirement to comply with NEPA. As discussed above, the NEPA process continues to evolve but the major steps are well defined. The first step is to determine who will act as the lead agency. In many cases this is clear from the start, such as a site located solely on land administered by a federal agency like the BLM. However, in some cases project components may involve land administered by multiple agencies, such as the BLM and U.S. Forest Service. In this case it is important to meet with all involved agencies to gain agreement on roles and responsibilities. In most cases the lead agency will require the project to fund their staff for the project as well as technical consultants hired to assist in preparation of the environmental document and supporting technical analyses. The project may already have retained technical consultants so it is important to work closely with the lead agency so that there is minimal duplication of effort.

Once the lead agency has determined that it has sufficient information to characterize the project, based on the project description, they will determine the appropriate level of NEPA documentation that is required. While some small projects may qualify for the less-intensive Environmental Assessment (EA) process, most utility scale solar projects will require an Environmental Impact Statement (EIS). The EIS process is initiated by the lead agency issuing a Notice of Intent (NOI) that is sent to other agencies who may have interest or jurisdiction and the public. Prior to the NOI being issued, the project should be actively engaged with other regulatory agencies as well as with potential NGO and public groups who may be interested or opposed. Early outreach is very important to avoiding unpleasant surprises later on in the process. You can be assured that the lead agency will receive comments from numerous agencies. The public will also get their say when the agency holds public meetings knowing as "scoping." The purpose of the scoping exercise is to determine the issues to be examined (i.e. the "scope") in the EIS. The project should be actively participating in the scoping process as it allows you to develop an awareness of the issues of concern, engage with potential opponents, and develop alternatives that can reduce the likelihood of strong opposition.

Once the scoping process has concluded, the lead agency will produce a scoping report outlining the issues and concerns identified in scoping, and those to be addressed in the EIS. At this point many projects make a significant error; the lead agency staff and their consultants go off to prepare the draft EIS and many projects become removed from the process. NEPA case law is clear that the EIS preparation needs to be an independent process but this does not preclude you from keeping open lines of communication, providing updates as the project design becomes more refined, and monitoring the assumptions being used in the technical analyses. These projects are large and complex so it is natural that the design and some features change over time. A good EIS must reflect as closely as possible the project to be built. Significant changes between the draft and final EIS can lead to legal challenges that the public was not allowed to comment on what is truly proposed, or a requirement by the lead agency to issue a supplemental draft document. Further required mitigation, if not managed, can overwhelm the project economics. There are usually opportunities to "negotiate" between mitigation options to select the best for all parties. Failing to take advantage of this will nearly always result in more project costs not less.

Another reason to be closely involved in the process is that the project is often in the best position to provide baseline environmental data and technical analyses related to certain issues. In addition, the project is also likely to be concurrently securing approvals or concurrence from other agencies (see Figure 3.2). Resource agencies such as the U.S. Fish and Wildlife Service have their own mandates, protocols for data collection, technical analyses, and format for documenting their findings. All involved agencies must be in close coordination and the project needs to be the facilitator of this. Staff of many agencies have multiple projects and may struggle to stay on schedule or to take the time to resolve potential conflicts with the lead agency. This can lead to serious schedule slippage and the potential for additional studies if not resolved early. For example, an air quality resource agency may desire that the project incorporate ponds on site during construction to allow for quick response dust control. However, the biological resource agencies may be concerned that such ponds will attract predatory birds that may prey on endangered species found on site. The project must play a key and ongoing role in identifying and resolving such potential conflicts. In addition to frequent meetings, a standing weekly call with all agencies invited can go a long way to addressing these concerns. Project permitting requires significant time on the road meeting and communicating.

Prior to and during preparation of the draft EIS, the project should be aggressively supporting the process by providing baseline environmental data and technical analyses. No one knows the project better than the project developer so we recommend that the project provide as much of the technical "heavy lifting" as the lead agency will allow. Baseline data, like surveys for cultural resources and biological resources, are subject to specific protocols that should be approved by both the lead agency and the resource agency prior to conducting the surveys. Be aware, it is not uncommon for different agencies to disagree on the appropriate survey technique. Failure to address such disagreements may result in the need to re-survey. Some surveys can only be done in certain seasons and no project wants a year delay while it waits for the survey window to open again.

Later in this chapter we will discuss the increasing role of Native American Tribes in the permitting of solar generation facilities in the desert southwest. It is important to note here that early consultation, both by the project and the lead agency, is important to the development of sound survey protocols for cultural resources. Many of the tribes who may be interested can be identified during the scoping process but we encourage ongoing outreach efforts as the project progresses.

After the baseline data has been compiled and the technical analyses are completed the lead agency produces the draft EIS. This is truly a significant milestone in the project development. For large solar projects these documents can be over 2000 pages in length when you include the supporting reports and technical analyses. Given the importance and scope of the EIS, the developer should be closely involved with the lead and cooperating agencies at every step. Prior to the public issuance of the draft EIS, the lead agency typically produces an internal review draft referred to as the "Administrative draft EIS." While some lead agencies may be resistant, the project developer should strive to review the Administrative draft. We believe that this is in the best interest of both the agencies and the project developers. If the draft EIS contains errors, it is very difficult to correct these without causing confusion with the public and other stakeholders.

The lead agency will make the draft EIS public, and will use the list of those who participated in scoping to ensure that interested parties are aware that the document is ready for review. Typical comment periods extend for 30–60 days but the duration of the public comment period is determined by the lead agency based on the perceived interest in the project. During the comment period the lead agency will hold workshops to receive public comment on the draft. As we will discuss later, it is not uncommon these days for solar project approvals to face legal challenges. Therefore, the lead agency will take care to document all comments so that the administrative record is complete. While scoping is for the purposes of determining the issues to be examined during the environmental review process, the comment period is intended to receive specific comments on what the agency proposes to approve. This is the stage in the project when opposition often becomes more strident. It is important for developers to continue to reach out to all stakeholders even when public comment becomes emotional and sometimes personal. Developers should also remain open to comments or suggestions that may actually improve the project. All of us like to feel as though our singular views are important, a project can generate significant goodwill by showing it is willing to take public feedback seriously.

Once the public comment period has closed, the lead agency will compile all the comments and determine what level of response is necessary. Agencies often rely on the project developer to provide technical support to their responses, having resources ready to respond quickly is critical to keeping on schedule. Any project changes that are being considered in response to ongoing design efforts should be discussed with the agency to see if they can be incorporated in the final EIS. This requires walking a fine line. If the agency feels the change is material they may move to issue a supplemental draft document. If this happens be prepared to watch longingly as your schedule flies out the window. However, the agency approval and subsequent conditions are highly detailed so you may have very little flexibility for project "tweaks" without being in violation of your approvals. As mentioned previously in this chapter, it is a bad idea to delay design until the permits are almost completed.

The agency will publish the final EIS and allow for public comment. It will then issue its Record of Decision, supported by conditions of approval. It is important that the timing of the NEPA process "flange-up" with other approvals from other agencies, as well as any other state and local reviews. In many cases you must also coordinate your NEPA process with the state and local processes.

Given how much "fun" the NEPA process is, many developers have favored privately owned sites with no federal nexus. In many states there exists a robust environmental review process that often looks like NEPA but has different requirements. Some of these differences can be significant and we strongly advise all developers to consult with legal counsel. There has been a push, especially in California, to combine efforts to produce an environmental document that meets the needs of all agencies. The results have been mixed. While we recommend discussing the potential for a combined document with the federal, state, and local agencies, it may be easier to produce two documents from the same basic information and technical analyses. This decision is highly project specific.

State and local processes can vary considerably, so we recommend retaining advisors or legal counsel experienced with similar projects in the geographic area. In some states the environmental review process will be led by a state agency and in other cases by a local agency such as the County Planning Department. It is not practicable to cover all the potential variations here but the overall process and key issues are similar those for the federal process. That is, the lead agency will produce a draft and final environmental document, involving the public at each stage. In many cases the lead agency will coordinate closely with others such as wildlife, air quality, and cultural resource agencies.

One final note on permitting for solar generation projects is a reminder that your project involves both construction and operation. Given the large amounts of land required, the potential for large construction workforces and long construction periods, much of the controversy in recent projects has centered on the construction period. A great irony is that many projects find themselves addressing greenhouse gas issues related to their construction fleets; something few, if any, of us anticipated. Therefore it is important that the construction plan included in your project description be designed with the environmental review process in mind. Things such as shuttles for workers, low-emission equipment, and noise reduction measures are all things to consider. Remember, in the end a solar development project is still a complex industrial facility and in many ways no different, from the developer's point of view, than any other power project.

3.5 THE ROLE OF THE NON-GOVERNMENTAL ORGANIZATION

As we've discussed, the permitting process is a very public one. As time has gone on, members of the public with common interests have come together to form organizations to get their message out and to protect things seen as valuable to the organization. NGOs such as the Sierra Club are well known but there are numerous organizations that have interests in geographic areas with strong solar resources ranging from the very small and thinly funded to the large and well-staffed. The ability to express views, solicit support, and raise funds via the Internet has facilitated the growth in the number of NGOs.

This impacts the development and permitting process in many ways. First is the fact that regulatory agencies give more weight to comments received from organizations than from an individual. That does not in any way imply that individual comments are ignored, it is merely a fact of life in the process. Second is that well-funded NGOs can retain technical and legal support to dispute the information in an environmental document or permit application. This is clearly something that the average individual cannot do, and it serves to force the agency to treat such an NGO with care and to make sure that their positions are considered. Recently, NGOs have become very active in the environmental review process for solar generation facilities. As of early 2011 there were at least six active lawsuits challenging federal and/or state approvals of solar projects. These kind of challenges could be dismissed, resulting in the need for supplemental environmental review, or force additional mitigation measures. Regardless, the cost and schedule delay associated with a legal challenge is significant.

It has gotten to the point where some developers just assume there will be a legal challenge and build it into their schedule. While that may be useful for worst-case planning, the goal of development should be to engage with all stakeholders on an early and frequent basis. You may not like what you hear from some NGOs but having them clearly communicate their position and concerns will allow you to address these concerns directly in your environmental document, or to incorporate design considerations that may make the project more acceptable. This does not mean you should redesign the project to gain their support. There are many NGOs with differing concerns, and you cannot please everybody unless you cancel the project. At least one recent project elected to sign "side agreements" with a few NGOs wherein they agreed to mitigation measures above and beyond those required by the permitting agencies. In return these NGOs agreed not to file a legal challenge to the project. Unfortunately, other NGOs remained unhappy and the project was challenged nonetheless.

The best path is to treat all stakeholders, both individuals and NGOs, with respect. Early and ongoing engagement allows for consideration of project features that address concerns, and also provides you with insight as to areas that may be challenged. It also demonstrates the project's commitment to a cooperative process. This is especially important when dealing with Native American Tribes and issues related to cultural resources. In the past, developers have allowed interaction with the tribes to be done solely by the lead agency under a regulatory consultation process designed to ensure that Native American concerns are included in the environmental review process. Recently, a number of the Native American groups have expressed the view that this formal consultation process is not adequate and have challenged some project approvals. Given the Native American's unique perspective on cultural resource issues, having an open and ongoing dialogue with them can avoid problems later on in the permitting process.

The bottom line is that stakeholder outreach is not just something that is nice to do. It is a critical part of a successful development.

3.6 SOME PARTING THOUGHTS

Many of us have spent decades developing generation and transmission projects. While we have tried to highlight recent issues that confront utility-scale solar generation, many key tenets of development continue to apply. First, and most important, is that the effort (notice I did not say money) spent up front to select a good site, doing enough design work to understand your project, and engaging early with the multitude of regulatory agencies who may have a role in permitting will pay off in the end. There is no free lunch in the development of a generation project, so commit to attending to the details right from the start.

Secondly, recognize that you cannot please everyone. While it is important to consider suggested changes in response to agency and stakeholder concerns, "gold-plating" your project will not guarantee your success. In fact, all it will do is make it more difficult to operate, and result in a lower rate of return. Keep your eye on the end goal of an approved project, attractive for financing, and ready for construction.

Lastly, recognize that the regulatory and policy frameworks are constantly changing. Take the time to monitor these developments and understand how they might impact your project. Strive to be proactive, not reactive.

Chapter 4

Development: Land

William Hugron, Jason Keller and Tyler M. Kropf

One might think that an immense, vacant, sunny field with no buildings or agriculture in sight would be a perfect canvas for solar development. There are many areas of concern one must look at in finding the ideal site for renewable energy development. This chapter will discuss some of the issues to consider when analyzing a piece of land for a solar facility. Although one can never be sure where unexpected hurdles may be found, prior knowledge of solar industry technology and development, as well as knowing what to expect with land entitlements, will help best prepare any user to find suitable land and get through the approval and construction processes.

In the late 1970s and into the early 1980s, NASA, the U.S. Department of Energy, and private energy groups began looking at the vast desert lands of the southwest United States as the perfect testing grounds for the large scale alternative energy source of solar energy. Solar development at that time covered areas of approximately 100 acres of land and supplied energy to several dozen homes. Today, the sun-drenched land of the southwest is again the land of opportunity for solar energy production. A lot has changed in the last 30 years. The advances in technology and a continued push by the public and private sectors have utility-scaled solar projects covering thousands and thousands of acres of land, and supplying clean renewable energy to hundreds of thousands of households. With the growing shift toward renewable energy and the increased demand for development of utility-scaled solar projects there is only one constant; not all land is created equal.

4.1 BRIEF OUTLINE OF LAND USE AND ENERGY CONSUMPTION

The United States has a total land area of nearly 2.3 billion acres. As of the newest published accounting by the USDA (United States Department of Agriculture) in 2007: nearly 671 million acres (30 percent) was in forest use, 614 million acres (27 percent) grassland pasture and range land, 408 million acres (18 percent) cropland, 313 million acres (14 percent) special uses (primarily parks and wildlife), 197 million acres (9 percent) for miscellaneous and other uses, and 61 million acres (3 percent) for urban land that is home to 75 percent of the population (Nickerson et al., 2007).

One of the largest concentrations for solar development has been the southwest United States and in particular, the state of California.

California has a total land area of approximately 104 million acres with agriculture utilizing nearly 43 million acres, wilderness area at 14 million acres (includes National Parks), Bureau of Land Management (BLM) over 15 million acres, Department of Defense over 3 million acres, and urban land at 3.5 million acres (USDA, 2012).

The Mojave Desert, with a size of over 16 million acres comprised of portions of California, Utah, and Nevada, is quickly becoming a focused location for solar energy development because of high sun exposure and large tracts of undeveloped land.

4.2   SOLAR MATH, ENERGY CONSUMPTION AND THE LAND PERSPECTIVE

Following are three tables depicting various units of measurement for land and energy consumption.

A watt measures the rate of energy conversion and it is the main unit of power used in photovoltaics (PV).

The current industry standard for PV technology is approximately 8 acres of land per 1 megawatt (MW) energy capacity. This takes into consideration setbacks for roads, access corridors in the main generation area, inverters and transformers, operations and maintenance facilities, fencing, and lighting.

Electrical energy is generally measured in kilowatt-hours (kWh). If a PV module produces 100 watts for 1 hour, it has produced 100 watt-hours or 0.1 kWh. PV modules are labeled with their peak power output. The peak power output is the maximum power (measured in watts) the panel can generate in standard test conditions including 1000 watts per square meter of solar radiation.

Consider the amount of land needed to harvest enough energy to provide for the energy consumption of the entire state of California. In 2010 alone, California consumed 250,384,000 megawatt-hours (MWh) (California Energy Commission, n.d.).

Using the sun exposure in the Mojave Desert region of California, which has been measured on average at 7.0 (kWh/m²/day) through the National Renewable Energy Laboratory's (NREL) Solar Prospector, the following is explored:

$$1 \text{ acre} = 4{,}046 \text{ square meters}$$
$$7 \text{ kWh/m}^2\text{/day} \times 4{,}046 \text{ square meters} = 28{,}332 \text{ kWh/acre/day}$$

Taking a 20% efficiency rating of a proposed solar panel for each acre of land would produce

$$5{,}665 \text{ kWh/acre/day or } 0.005665 \text{ MkWh/acre/day}$$
$$365 \text{ days/year} \times 5{,}665 \text{ kWh/acre/day} = 2{,}067{,}725 \text{ kWh/acre/year}$$

California Consumption (2010) $\dfrac{250{,}384 \text{ MkWh}}{2.067 \text{ MkWh/acre}} = 121{,}091$ acres

Hence, all of California's electricity can be produced from approximately 121,091 acres, just less than 200 square miles of sunshine. Lake Mead, behind Hoover Dam, covers more than 200 square miles. A solar array the size of Lake Mead would provide enough energy to make California energy independent!

The growing popularity of renewable development is evident with at least twenty-nine states passing renewable electricity mandates, six states renewable energy goals, and one state with an alternative energy mandate. No matter what one might call it,

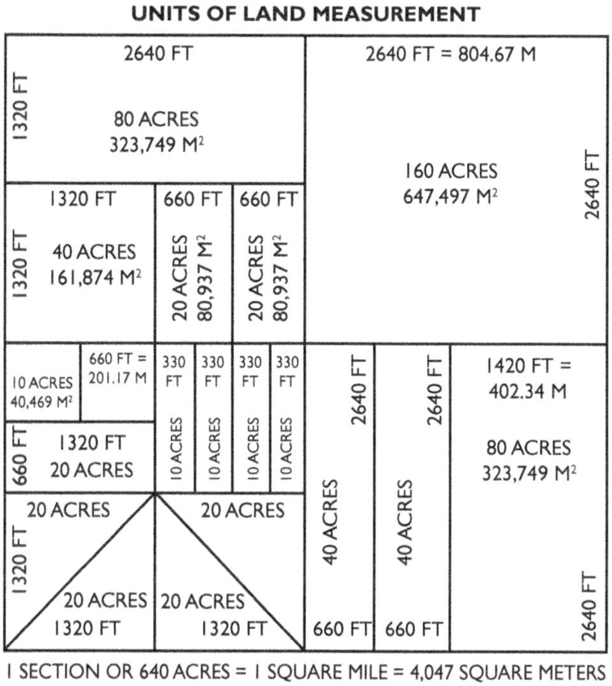

UNITS OF LAND MEASUREMENT

2640 FT	2640 FT = 804.67 M

1320 FT

80 ACRES
323,749 M²

160 ACRES
647,497 M²

2640 FT

| 1320 FT | 660 FT | 660 FT |

1320 FT

40 ACRES
161,874 M²

20 ACRES
80,937 M²

20 ACRES
80,937 M²

660 FT = 201.17 M

10 ACRES
40,469 M²

| 330 FT | 330 FT | 330 FT | 330 FT |

2640 FT

2640 FT

1420 FT =
402.34 M

80 ACRES
323,749 M²

660 FT

1320 FT
20 ACRES

10 ACRES
10 ACRES
10 ACRES
10 ACRES

1320 FT

20 ACRES

20 ACRES

40 ACRES

40 ACRES

20 ACRES | 20 ACRES
1320 FT | 1320 FT

660 FT | 660 FT

2640 FT

1 SECTION OR 640 ACRES = 1 SQUARE MILE = 4,047 SQUARE METERS

Figure 4.1 Units of land measurement.

the push for utility-scaled renewable energy development is growing, and the need for land and sites to build these projects is increasing.

California has an aggressive approach with a Renewable Portfolio Standard (RPS) that mandates California utilities and other electricity providers have until 2020 to draw 33 percent of their power from solar panels, windmills and other renewable sources. Looking at the 2020 mandate requiring 33 percent of power to come from renewables, and using solar modules as the technology used to achieve this, approximately 50,000 acres of land would be needed. This would generate enough energy for approximately 100,000,000 homes. Although this seems like an incredibly large amount of land, when put into perspective protected land is already set aside in the California desert for other uses many people are much less aware of than say for solar energy. Let us explore this in the following figure:

Land use in California desert region.

Land use	Acres
Desert Tortoise Preservation	4.8 Million
Mojave Ground Squirrel Preservation	1.7 Million
Defense Department Preservation	3.3 Million
Off Highway Vehicles	0.7 Million
Renewable Generation (2020 goal 33%)	0.05 Million

Source: Bureau of Land Management.

Table of Land Measurements	
LINEAR MEASURE	**SQUARE MEASURE**
1 inch = 0.0833 foot	1 feet^2 = 144 inch^2
1 foot = 12 inches	1 yard^2 = 9 feet^2
1 yard = 3 feet	1 acre = 4,840 yards^2
1 mile = 1,760 yards	1 acre = 43,560 ft^2
1 mile = 5,280 feet	1 mile^2 = 640 acres
1 yard = 0.0914 meters	1 mile^2 = 1 section
1 mile = 1,609.34 meters	1 mile^2 = 4,046.856 meters^2

Figure 4.2 Table of land measurements.

Table 4.1 Table of energy output per unit area.

1 standard panel	15 square feet (varies by manufacturer)
1 square foot of standard solar panel	13 watts of energy production
1 acre	4,047 square meters
1 acre on average can support	125,000 watts = 125 kW = 0.125 MW

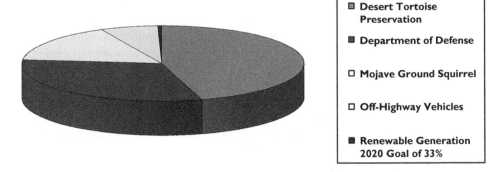

- ▣ **Desert Tortoise Preservation**
- ■ **Department of Defense**
- ☐ **Mojave Ground Squirrel**
- ☐ **Off-Highway Vehicles**
- ■ **Renewable Generation 2020 Goal of 33%**

Figure 4.3 A comparison of key land uses categories in California.

If solar energy ever became the country's primary energy source would there be a bluish panel at every corner of every city or on every horizon throughout the country? Contrary to popular opinion, relying on solar energy would offer a landscape almost indistinguishable from the landscape known today. What is important is that land development be done in an ecologically efficient and responsible manner.

4.3   PREPARATION PRIOR TO SEARCH

There are tremendous efforts taken on research, analysis, and studies for the manufacturing and analysis of solar technologies. Each technology is designed to be proficient under specific environmental conditions in order to maximize efficiency and output. The job of a land advisor is to take project requirements and find the best possible site scenario to maximize the potential output of each respective technology.

Utility-scale developers, owner/operators "end users" of energy, and will provide the most extensive and demanding set of requirements for the procurement of viable sites for solar development. They will be running their business at this site and will be most impacted by the resources, conditions, and jurisdictions of that location.

A seasoned solar developer understands that each region, county, and city as well as utility coverage areas can greatly alter the time, cost, and efficiency of a solar development project. What all developers strive to locate is a site that will enable them to construct a solar project in the shortest amount of time, with the least resistance from government agencies and the public, and at the lowest cost. The varying challenges of regional jurisdictions should be balanced with specific technology requirements in compiling information that will be used specifically for the analysis of prospective land sites. Whether there is an in-house project manager, a land acquisition executive, or an outside land advisor or broker, these pre-qualifications or land specific requirements if identified up front, will save significant time and resources.

There are certain criteria that must be identified by any developer prior to even beginning the search for land. Using this information will endow a land advisor with the ability to take a broad approach and hone in on the most ideal land sites.

Clarifications of the following items are necessary prior to beginning a search for land:

a) Land Size (acres)
b) State-County-Region
c) Utility Provider Preference (Southern California Edison, PG&E etc.)
d) Project Size (MW)
e) Distance to Interconnection
f) Proximity to Substation
g) Topography
h) Water requirements (if any)
i) Direct Normal Irradiance (solar irradiation)
j) Privately owned or not

4.3.1   Land size

Most utility-scale projects (projects that will tie into the transmission grid) will vary in the broad range of 1 MW to 100 MW. As in most land development, the size and complexity of a project will determine the amount of approvals, permits, and conditions required.

4.3.2 State-County-Region

The specific area chosen for development will also determine how many jurisdictions a project will fall under. A property in the city may fall under city, county, and state jurisdiction. Solar developers and end users want to be in the areas with local and state government policies that encourage renewable energy development and that have strong public support. These two factors have been shown to reduce or eliminate major investment risk factors including lengthy permitting processes and public opposition.

4.3.3 Utility provider

Each utility provider is unique and has separate requirements for interconnection. Many developers will determine where they want to develop based on the incentive or 'fast-track' programs sometimes offered through the utility companies.

4.3.4 Project size

Ground mounted projects that are currently deployed, as well as those in the planning and permitting stage, can be as small as 10 acres to as large as 3,000 acres. The size of the project can have direct impact on the amount of time it takes to entitle the project for development. As the project's size increases developers may look at developing a project in phases.

4.3.5 Distance to interconnection and proximity to substation

The site's proximity to a substation, as well as transmission or distribution lines, can make a significant impact on the final cost of site development, and may well contribute to a loss in energy efficiency. For every mile separating a parcel of land from the substation or to the delivery point, approximately 0.5 percent of energy can be lost.

It has been estimated that installation of transmission lines between 115–230 kilovolts (kV) for 1 mile will cost, at minimum, an additional $940 thousand to $1.1 million to the project, as well as additional permitting and time allocation for approvals (Ng, 2009).

Ideally, a parcel of land within 0.25 miles of a substation that has transmission or distribution lines running along the border of the property is what a developer will look for.

4.3.6 Topography

Whether constructing a concentrating solar power (CSP) or PV site, the topography is important to the overall cost and viability. The greater a necessity it is to grade the site, more necessary in CSP projects, the more prohibitive the site will become for development.

4.3.7 Water requirements

Each technology is different with regards to water requirements. For a CSP site, hundreds of acre-feet of water may be required in order for cooling to function properly for electricity generation. An acre-foot is equal to one foot of water over a one-acre area. If water is not available on the site it can be purchased or leased from owners within the adjacent water district. Water isn't necessarily required on site for PV technology since trucks can be brought in a few times a year to wash the panels, which will provide for a larger potential number of conducive sites as water rights are often not associated with surface rights. One gallon of water is typical used per cleaning of 1 panel in a plant, with several thousand panels in a large project this can quickly add up.

4.3.8   Solar radiation

Solar radiation is defined typically in (kWh/m²/day) kilowatt-hours per square meter per day. Scattering, absorption and reflection that can impact radiation levels based on the location of land. Most solar energy systems, such as PV and CSP look for levels of 6.0 kWh/m²/day and above, with CSP needing direct sunlight. This should be the first step in analyzing a region or area of land. Minimum solar radiation levels can have a huge if not fatal impact on the technologies being used. Having simply 10 percent better solar radiation from one site to another provides much flexibility in its development; power purchase agreement (PPA) prices can be more competitive with a site that carries stronger solar resources since more energy is being generated.

4.4   USING TECHNOLOGY IN SITE SELECTION PROCESS

The use of technology to find and analyze land saves time and expense. The search will always begin with a desktop analysis. The ability to be two or three places at one time has been made possible due to the advancements in computer mapping programs. Only after identifying a site that passes all portions of a desktop analysis, can a site visit be justified. Solar developers can make the best use of their resources by identifying as many potential sites as possible in a particular region and narrowing them down through armchair analysis. This task is accomplished by utilizing several available online tools.

The most important tools at the disposal of any developer are mapping programs and aerial photography. With online mapping, data is continuously updated and the user can switch between street maps and aerial views with one click of a button. There are many different mapping programs on the market with varying capabilities. Several of the most comprehensive mapping and solar data sites will be discussed.

Anyone with access to the Internet can use Google Earth and view real estate in many parts of the world at no cost. A user can type in an address and fly instantly to the property, save data from a previous search, apply layers, and overlay various images. One function that Google Earth offers that is a very valuable resource for land acquisition is the Google Street View program. Since Google Street View was fully implemented into Google Earth in 2008, it has allowed a user to not only observe a satellite image from above but also view the surroundings with 360-degree views. The program covers the majority of public paved streets in the United States, although there are some remote areas where access is limited. Using Google Street View, a user can be sitting at an office in Newport Beach, California and be analyzing a parcel of land in Phoenix, Arizona. This is the next best thing to actually being there in person. From this vantage point, the user can visualize power lines and see where they intersect the subject parcel, which is not always possible from a one dimensional satellite view. Also, by looking at the site from ground level, you can get a real sense of the scale of the topography that may not be as easily discernible with a topographical map.

Another program that provides a user the ability to see land from various vantage points is Bing Maps. With Bing Maps a user can fly to any location just as with Google Earth, but what is unique to Bing Maps is the Bird's Eye View technology. Just as the name implies, once a property is located, the user can access imagery of properties at 45-degree angles taken by low flying planes. Properties can be viewed from north, south, east or west. These angled views allow an aerial view with better depth perception and clarity than that of a satellite image. With the technologies of Google Earth and Bing Maps, land can be studied from many different angles to help establish its viability as a solar site.

A third program ACME Mapper allows the user to switch between street map, aerial image, topographical map, and even weather surveillance mapping while maintaining the same location. Currently, online topographical maps produced by the U.S. Geological Survey (USGS) can be challenging if viewed by themselves, due to their lack of identification layers. ACME Mapper provides the ability to toggle back and forth with an aerial image or street map, making it easier to identify exact reference points on the map. The weather surveillance mapping contained within ACME Mapper is called NEXRAD, which stands for Next-Generation Radar, and consists of a network of Doppler weather radars operated by the National Weather Service. NEXRAD mapping is used to track precipitation and atmospheric movement. If an area is known for its heavy amount of precipitation it should obviously be avoided for any solar development.

Being able to see what potential sites look like is paramount to the success of any desktop analysis. There are several other parameters that must be fulfilled before viewing potential solar sites using Google Earth or Bing Maps. One such parameter is the type of solar energy required. There are two types of solar radiation, direct normal irradiance (DNI) and global horizontal irradiance (GHI). DNI is described as the amount of solar radiation received perpendicularly by a surface that comes directly from the sun in a straight line (versus diffuse light like the light available under a tree, there is shade and hence not direct light but there is still ambient light) at any time throughout the day. CSP technology relies solely on DNI radiation. GHI includes DNI, but it also measures the radiation that comes from scattered or reflected beams of sunlight. For PV technology, the measure of GHI radiation is used. An excellent resource to verify solar radiation in any given region in the United States is NREL's website which provides the user with mapping tools to view the radiation levels for both DNI and GHI.

Another important feature of NREL's website is the slope calculation of a selected landscape. Solar developments typically require relatively level ground. While PV projects can be more forgiving in this respect, CSP projects require nearly laser-level flat landscapes with the ideal being less than 1% slope. The desired level of slope can be selected, allowing the user to rule out adverse areas.

Counties and cities are incorporating online mapping using geographic information system (GIS) capability into their websites, allowing users to view General Plan data, zoning data, and other forms of data about the selected region specific to that county. Being able to access such information from various municipalities rather than using the traditional method of phone calls can save a lot of time.

Probably the most comprehensive mapping site is available, for a fee, through Digital Map Products and is called LandVision. LandVision provides a user with tools to expedite the site procurement process. With this one program you can cover nearly all of the preliminary aspects of the desktop analysis. While Google Earth and Bing Maps are limited to searching for properties using a physical address, LandVision gives the user the capability to search by assessor's parcel number or APN (this is a property's unique and individual tax identification number), by street intersection, by zip code, by latitude/longitude, by city or county, by ownership name, by property information such as building size or year built (if there are improvements), by property sale information, and by lot size. While the program would be well worth having for this functionality alone, it offers much more as well.

In order to have the most current parcel identification and boundary data available, Digital Map Products contracts with most counties throughout the country. Various data sets or layers are overlaid onto satellite aerial imagery. The boundary data is updated as the counties update their data. The parcel overlays appear on the aerials from a zoom distance as far as 5.25 miles above ground level while still giving an expansive view of the landscape. In addition, when a given parcel is selected, the parcel boundary line becomes highlighted and the property information appears to the left of the image. From the displayed property information the user can access a property profile, plat map, view transaction history, or click a link for a business name lookup if the ownership is a limited liability company (LLC) or a Corporation. Each of these functions opens up in separate windows from the main screen. All property and title information is obtained through First American Title Corporation.

Depending on a specific project requirement it may become necessary to gather several smaller parcels of land together to create a larger land mass. Through the use of LandVision's user tools, the user can measure distances between points, calculate the area of any parcel or groups of parcels, draw polygons, as well as line and circle to focus on key parcels and add descriptive labels and symbols. All data can be saved as projects within the LandVision program for future use.

LandVision also incorporates many layers that can be displayed individually or several at a time. Layers such as transportation labels, census tracts and blocks, school districts, city, county and zip code labels, and postal codes aid the user in basic location identification. Layering transportation labels with parcel boundary information will supply very pertinent information, for instance, if there is direct road access to the subject property, a factor that is not always clear from a mapping program separate from parcel boundary data.

There are also more advanced layers for data such as county land use categories, water sources, government land, California Williamson Act, and natural hazard zones that are crucial to be familiar with for any development. With the county land use layer, property types such as vacant land can be isolated so that the search results will only return properties that fall within the specified land use category. If it is possible to determine whether a property lies within a flood plain or within a natural hazard zone in one glance the user will know what areas to avoid and can move on to the next area unburdened by these potential building obstacles.

Moreover, it is possible to import additional layers into LandVision in the form of shape files to streamline a search even more. By importing layers such as power lines and substations, which can be purchased by region through several organizations who track this data, it is possible to visualize precisely how a subject parcel lays out against power lines, determine the power capacity and calculate the distance to the nearest substation. Imagine the benefit of using one mapping program to view a map of an entire county showing all areas to avoid based on the advanced layers mentioned above along with views of power line and substation locations. With this powerful information, the user can pinpoint the address required to contact landowners for site acquisition. This is the invaluable resource LandVision provides. By repeating this process over and over, many potential development sites can be sourced.

Like any power project, developing a solar energy project is a complex process. For a solar energy project to be successful, the many parts of the development process must be layered and resolved. Thankfully, there is technology to aid in these endeavors.

4.5   BLM LAND OVERVIEW

Many living in California or the Western states are familiar with a lonely desert stretch of Interstate 15 that connects Nevada and California. Most everyone that makes that drive comes to a quick conclusion that there is plenty of room for expansive building and development. The one thing many do not realize as they see mile after mile, hour after hour of the same open desert highland is that the majority of the land is public, owned by the BLM, and in its current state cannot be developed. To illustrate this, the BLM map in Figure 4.4 highlights the vast amount of public land along the stretch of Interstate 15 that extends from San Bernardino County, California to Clark County, Nevada.

The question then may be posed, if this land is to remain undeveloped, why can't a portion be utilized for renewable energy generation? What is the difference between developing solar on public lands versus private lands? There are several advantages and disadvantages to developing a solar project on these two types of land ownership.

4.5.1   BLM vs. Private land

Even with a new, more systematic approach for developing on public lands and taking into consideration the recent momentum of the first ever approvals for solar development with the BLM, a developer will have greater flexibility and control of a project by devoting more time and resources to private land versus public land development.

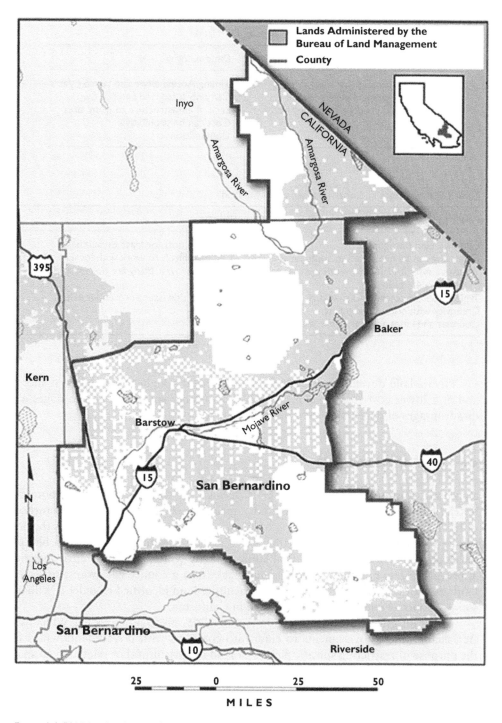

Figure 4.4 BLM land in Inyo & San Bernardino Counties (Courtesy of U.S. Department of Interiors, BLM – Barstow field office).

Table 4.2 Advantages and disadvantages of BLM land.

BLM Public Land: Advantages	Disadvantages
Fixed cost-rental program	Timing. Access alone can take 5+ years
One lead agency. Experienced.	Can only lease land and not own
Standardized process. Becoming faster.	Lack of infrastructure to many sites
	Cost can be prohibitive
	Competition

Table 4.3 Advantages and disadvantages of private land.

Private Land: Advantages	Disadvantages
Can own land outright	Unknown cost for lease or purchase
Permitting can be expedited. In some unique cases a typical 12 to 18 month process can take 2 to 4 months.	Can be difficult to work with local jurisdictions for approvals. Many are not experienced.
Typically better access, infrastructure	May have title issues preventing sale
Creativity with pricing and terms	
Counties and cities advocating for you.	

Private land development, thus far, has provided a higher probability of success and at a lower cost to a developer. Table 4.3 outlines some of the advantages and disadvantages of developing land in both sectors.

4.6   WILLIAMSON ACT

Significant portions of sites that make it through the filtering process for potential development are Agriculture Zoned lands. Agriculture land is considered "disturbed" or has been unsettled by the plowing of land or growing of crops that lowers the possibility of animal species from settling on it. This is a double edged sword because developing on disturbed agricultural land normally will expedite the environmental approval process but may also cause pushback from a county who wants to see the land used for agriculture. Much of the agriculture land identified is under a Williamson Act contract exclusive to sites located in California.

The California Land Conservation Act of 1965, also known as the Williamson Act, enables local governments to enter into contracts with private landowners for the purpose of restricting specific parcels of land to agricultural or related open space use. In return, landowners receive lower property tax assessments. It is estimated that the Williamson Act can save agricultural landowners from 20 to 75 percent in property tax liability each year, without this deduction many would not be able to afford to farm. Under the Williamson Act, an owner of agricultural land may enter into a contract with the county if the landowner agrees to restrict use of the land to the production of commercial crops for a term of not less than ten years. The term of the contract is automatically extended each term unless notice of cancellation or nonrenewal is given.

The Department Of Conservation, Division of Land Resource Protection provides three circumstances a solar generation facility may be approved by a city or county:

1) Compatibility

   a) The conditional use permit requires mitigation or avoidance of on-site and off-site impacts to agricultural operations.
   b) The productive capability of the subject land has been considered as well as the extent to which the solar power generation facility may displace or impair agricultural operations.
   c) The solar power generation facility is consistent with the purposes of the Williamson Act; to preserve agricultural and open space land, or support the continuation of agricultural uses, or the use or conservation of natural resources on the contracted parcel, or on other parcels in the agricultural preserve.
   d) The solar power generation facility does not include a residential subdivision.

   There are a host of factors for a county or city to weigh and consider in making the above required findings. These include but are not limited to: the availability of irrigation water, size of the solar power generation facility, size of the contracted parcel, slope, placement and location of solar panels, and types of mitigation and avoidance offered. Because each situation is so fact specific, the Department stands ready to assist cities and counties in performing the required compatibility analysis.

2) Non-renewal
   Williamson Act contracts may be administratively or unilaterally "non-renewed" either by the landowner, the city, or the county. Non-renewal of a Williamson Act contract starts a ten-year process toward the contract's expiration, during which property taxes will be returned to their full amount.

3) Cancellation
   As an alternative to non-renewal, the landowner may seek to immediately cancel the Williamson Act subject to discretionary approval by the local agency having jurisdiction over the contract. Landowners who cancel Williamson Act contracts are required to pay a fee of 12.5% of the unrestricted value of the property to the State. (Department of Conservation, 2011).

The contract language of the Williamson Act is void of specific language pertaining to solar facility development as it was written prior to the mainstream solar industry. Counties have taken liberties to rule on what the Williamson Act would have allowed for had the solar industry been around at the time of its inception, in order to promote renewable development on these lands. When identifying a potential parcel of land with a view to fulfilling the regulations of the Williamson Act it is important to understand the current standards and policies of the county regarding compliance, non-renewal and cancellation, as well as the timing and cost associated with each of these possibilities.

4.7 LEASING, PURCHASING AND OPTIONS

The question often arises; is it better to lease or purchase land for a renewable energy development? There are several items to consider. One positive element of leasing is there will be less startup capital to contend with. Only having to make monthly payments to a landowner allows a developer to keep the monthly expenses low and save their resources for project expenditures such as permitting costs, or the freedom to fund more than one leased venture with the same amount of funds necessary to purchase just one site.

As for purchasing, the fact is that what can seem like an enormous sum to pay at the close of an escrow is oftentimes eclipsed by the total amount of capital spent over the life of a leased project. Not to mention, as owner of the property, a developer has more control and fewer restrictions. When a property is purchased, a developer does not have to seek approvals from, or answer to, a landlord for any reason.

Another aspect of purchasing vs. leasing, not often considered by developers, is that it is possible to obtain title insurance on 30+ year leases, which many solar leases are. With title insurance, a solar development could be sold as a leased investment. This creative strategy allows developers to complete a project then sell the land to an investor at a profit, retaining the leasehold interest while continuing to yield revenue from the energy sales. This is a tremendous exit strategy as well, should a developer who purchases a site decide they do not want to own any longer.

The answer to the question of leasing versus purchasing will depend on the goals and objectives of the developer. Often, the economics of the project may drive the decision to purchase or lease the property. Is the intent to sell a project at some point during the entitlement process, or to take the project all the way through development to later become the project owner and operator selling energy to the utility company? If the former were true, then leasing would seem the better alternative, minimizing the obligation to the land. If the latter is more accurate, then perhaps purchasing makes more sense.

Leasing property with options is another very beneficial tool for a developer. For the short-term developer, a lease with an option to purchase will be a marketable item to the eventual tenant who may want to exercise the option and purchase the property. For a project owner-operator, if a lease has been established for a few years and the property financing is in place, it becomes economically feasible to purchase. While negotiating a lease, options to purchase can be included so that after the lease commences the developer will have the flexibility to convert that lease into a sale under the option. All terms of the option must be established during negotiations.

There are a several items to consider on options as part of a lease:

- Option term length
- Cost to exercise the option
- Is the option transferrable?
- Can the option money be applied to the purchase price?
- Is the option money refundable?
- Is any portion of the lease payment applicable to the purchase price?

- Is the purchase price set at the time the option is established or will developer pay market price at the time the option is exercised?

Lease options allow the developer more choices that are extremely helpful during the development phase of a solar project. If a developer were to determine that continuing a lease would be the best course of action, then the developer would not be obligated to exercise the option and the lease could continue uninterrupted.

Options can also be used outside of a lease structure to secure a position for a future purchase of property. An option in this case allows control of a property at a minimal cost. While an option is in place the property is off the market to any other would-be buyers. During an option, a developer can conduct due diligence, secure project financing, process entitlements, or locate another buyer. The same considerations previously stated for leasing with options would apply, although the option money itself would be non-refundable since the land owner would not have the benefit of lease payments.

4.8 CONTRACTS

In dealing with both purchasing and leasing there are two types of contracts to consider: standard form and attorney drafted. Both standard and attorney drafted contracts are technically drafted by an attorney, however, standard form contracts are pre-drafted using industry standard language that would cater to a wide audience for varying transactions. Conversely, attorney drafted contracts are written for specific transactions. For straightforward transactions standard form contracts are recommended, as they will save time and expense. For transactions in California, Arizona, Nevada, Florida and several other states, consider using standard forms produced by the AIR Commercial Real Estate Association. These forms are regularly reviewed and updated by real estate attorneys to ensure compliance with state laws and regulations, which can vary from state to state, not to mention they are typically fair to all parties involved. Be advised when using standard form contracts, it is not recommended to enter into contract negotiations without the assistance of a licensed real estate professional. Additionally, it is highly recommended to have legal counsel review any contract prior to signing.

On the other hand, for transactions where there may be concerns such as property issues, environmental issues, title issues, water or mineral rights issues, and tax ramifications; consider hiring a real estate attorney to draft a contract to address these specific items. When there are multiple articles of information to address, the extra time and cost will be particularly worth the effort. With that said, if there are only a few issues, it may also be possible to hire an attorney to simply add supplemental language to a standard form contract.

Prior to entering into a legally binding contract, whether it be a Purchase and Sale Agreement or lease, it is prudent in many cases for a developer to submit a non-binding letter of intent outlining the main deal points around which the binding contract will be based. The deal points will include the names of buyer and seller, correct property description, option period terms, due diligence periods, and purchase price or lease rate over the course of the lease term. For a purchase, include escrow period, closing date, and down payment amount. Also, specify if the sale is to be an all cash transaction or if there will be financing involved. If choosing a lease, state the lease commencement date, occupancy date, and any options to extend the lease following the lease expiration. Once these main deal points have been agreed upon between all principals, the legally binding contract can be drafted. The following is an example contract:

[{Month/Day/Year}]
[{Lessor/Broker Name}]
[{Lessor Address}]
[{City}], CA
[{Zip Code}]

RE: ACRES: Approximate +/–00 Acres
     APN: 000-000-000
     ABC County, California

Dear Sir/Madam:

On behalf of **ABC Buyer** (Hereinafter "Buyer"), we are pleased to submit this Non-Binding Letter of Intent to Purchase land at the above referenced site based upon the following terms:

BUYER/DBA:	4.8.1.1.1 ABC Buyer
SELLER:	**XYZ Seller**
USE:	Solar Farm
PREMISES:	APN: 000-000-000 consisting of approximately 00 acres identified as parcel located in ABC County, California, together with all easements, rights and appurtenances thereto (collectively, the "Property").
PURCHASE PRICE:	$000,000, or $0,000 per acre, All Cash
ESCROW PERIOD:	Upon executing this Letter of Intent, Buyer shall draft a Purchase Agreement based upon the terms of this agreement. Seller agrees not to enter into any agreement nor market the property to any other parties before or during the Escrow Period.
	Escrow Period shall be a 24-month time frame during which time Buyer shall perform all Due Diligence. Escrow shall close [{Month/Day/Year}] or sooner. Buyer at any time during Escrow Period can give 30-day cancellation notice to close escrow or cancel this transaction.
	Buyer and its representatives will have access to the property during Escrow Period. Seller shall comply with Buyer in a timely fashion in order to obtain any necessary documents for the Due Diligence Purposes.
	The Buyer may be required to obtain entitlements, permits, and rezoning to develop a solar farm as part of this transaction. During this time the Seller shall not be required to pay for any development related costs.

For Due Diligence purposes, Seller shall cooperate and provide any documentation pertaining to the land in their possession.

OPTION PAYMENT: During Escrow Period, Buyer to pay a non-refundable monthly due diligence option fee of $0,000.00 per month paid monthly during the Escrow Period. Option payments shall cease upon close of escrow or upon cancellation of this transaction.

Seller to cooperate to the best of their ability and provide any and all information including studies, reports or tests pertaining to the property in a timely fashion.

INSURANCE: Buyer shall be responsible for maintaining insurance on the property and naming Seller as additional insured during Escrow Period.

CONFIDENTI-ALITY: Seller agrees to keep this Letter of Intent strictly confidential (including the identity of the Buyer and the proposed terms of the Letter of Intent and Purchase Agreement).

Please note that this is a non-binding letter of intent and no obligations between the parties are created as a result of this letter. Buyer and Seller understand that agreement to this letter of intent will not necessarily result in a fully executed purchase agreement. This proposal shall expire on [{*Month/Day/Year*}] at 5:00 PM Pacific Standard Time.

Buyer: **ABC Buyer**
By:
Date:
Seller: **XYZ Seller**
By:
Date:

4.9 PRICING

Obtaining pricing for land is not as simple as running a comparative market analysis for a tract of homes where homes are abundant and sales occur on a frequent basis. It takes time to research an area, analyze any comparable sales, and speak with local brokers as well as with property owners. While comparable, sales are normally what determine market pricing but this data is not always available, particularly in remote areas where solar renewable projects are more likely to be built. Sometimes, even if comparable sales are found, the properties are not true comparisons to a site that a group may be investigating. If researching a 400-acre site, for example, and the only comparable sales to be found are a handful of five to ten acre parcels, this information cannot be used as the basis for pricing on the larger site. Repeatedly property owners prove to be most valuable assets in determining value in cases where no viable, comparable sales data can be obtained. After running the desktop analyses, and selecting specific areas ideally suited for solar developments, the next step should be contacting the surrounding owners. The local residents usually bring considerable insight to a given area, having the advantage of prior knowledge of town history. Market value for land in these cases becomes what the owners are willing to sell it for. At this point the project developer must determine if a property owner's asking price can be justified based on the comparable properties transaction history and any other area data that can be found.

4.10   CONCLUSION

This chapter began by discussing the ample supply of sun drenched open land in the southwest United States and continued with specifics considering why not every piece of land carries the same potential for solar development.

   If a developer were to inquire about solar development with government agencies several years ago they would have been met with limited ability to progress due to antiquated, or non-existent, systems and procedures for the various types of solar applications. In a few short years not only are more and more jurisdictions prepared with policies and procedures for development, but also these same county agencies are also going out and marketing some of their land as "fast track" areas. These areas have already passed in-house engineering and environmental studies for expedited solar power plant development. Solar development on both private and public lands is becoming more and more standardized. Developers will be more adept at identifying a successful site for utility-scale solar development if they keep in mind that each parcel of land is unique, and are armed with a multi-faceted understanding of the solar industry's singular technologies and challenges.

Chapter 5

Development: Transmission

Arturo Alvarez, Jesse Tippett and Albie Fong

Developments for commercial projects such as shopping malls or housing developments evaluate the proximity and quality of key resources such as demographics and the local transportation support infrastructure. In energy projects, the second most vital characteristic after the evaluation of the energy resource (solar radiation in the case of solar energy) is the critical support infrastructure of the power transmission and distribution lines available to the project. Without reasonable access to transmission infrastructure, power projects cannot effectively deliver their power to the eventual customers. In this chapter, we will show the reader the core ideas regarding the process for evaluating, defining and ultimately implementing a plan for connection to the electrical grid to transmit power for a project based on industry examples.

In solar project development it is necessary to fully understand the transmission viability of a given project site at the very early stages of evaluation, and far in advance of gaining official site control. After determining the ultimate transmission capacity of the site, the developer can maximize the design potential of the site for power purchase agreement (PPA) and permitting purposes.

The electric interconnection process is associated with the physical connection of an energy generating project to the electrical transmission network. Electric transmission, often confused with the interconnection process, is the physical transfer of electricity from the power plant to the end user. Electric transmission includes the electric transmission lines, also known as power lines, and the grid operators that monitor the safety of this transmission of energy. Between the power plant and the end users there are several key electrical equipment pieces that allow the safe use of power that should be understood before we begin. They are step up transformers, substations, and step down transformers. A step up transformer is a piece of equipment that is used to take the power plant's generated power from low voltage and turn it into high voltage; this is done in order to incur minimal losses during the transmission of power from one place to another along lengthy distances of power lines. Substations house the transformers and safety equipment that is required to operate the grid. These are vitally important to the transmission system because they allow for power to be generated, delivered, and used between different regions of the grid and ultimately safely in our homes. At times, power plants utilize an existing substation where a step up transformer takes the power brought in from the plant and converts it into high voltage that matches the power line rating, prior to sending the power to the grid, which uses a series of equipment for safety and reliability purposes. Once the power is carried out of the power plant to the substation and placed on the grid, there is one last step that the power goes through before it reaches our homes. This last step reverses what the step up transformer did, and brings the power back to a low voltage that can be distributed to our homes. This final step is completed by another substation that houses a series of step down transformers that converts power to a lower usable voltage.

Interconnection is the physical aspect of plugging a new project into the electrical transmission grid. This process is governed by the interconnection permitting process, which will vary depending on a project's specific parameters such as nameplate capacity of the project, project location, and the entity processing the interconnection request for any given project. Ultimately, each utility establishes their own procedures and tariffs with their own interconnection agreements even with state laws to govern what is allowable under interconnection policies. The nameplate capacity of the project, how much power will it be rated to produce in megawatts, is the characteristic that will have the largest effect on the type and complexity of the interconnection process the project must go through. Typically, a project with a low power output rating or nameplate capacity will have a less complex interconnection permitting process than a larger project simply because less energy will need to be transmitted.

In Figure 5.1, a simplified electrical power transmission diagram, we see that there are differently squares that outline the different "boundaries" of a project. The upper left boundary would represent the generator with its step up transformer or substation, the upper right boundary represents the electrical grid, the bottom right boundary represents the substation, and the bottom left boundary represents the distribution lines and end users. All of these parts work together and with the oversight of the grid operators, the power is generated, dispatched, delivered, and used every day.

Locations with existing transmission lines and available transmission capacity, without requiring significant system upgrades are some of the most continuously

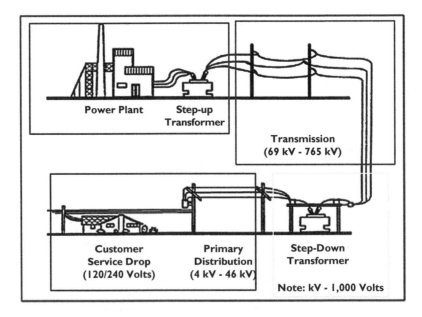

Figure 5.1 Simplified electrical system (Public Service Commission of Wisconsin).

sought after characteristics for successful renewable energy project areas. Depending on the project location, some projects may have multiple ways of interconnecting to the electrical grid; there may be multiple power lines crossing, adjacent, or in the nearby vicinity of a potential project. Existing transmission lines provide paths for the energy generated to reach the electrical grid and ultimately the end users. When transmission infrastructure upgrades are not required, this provides an economical advantage making the project more economically viable due to lower overall project costs. While requiring no additional cost for interconnection upgrade is preferred, this does not imply that a project requiring transmission infrastructure to be built or upgraded will not succeed. A project can still be developed without pre-existing transmission lines. However, it is important to factor in the added project cost of constructing transmission lines or upgrading transmission infrastructure. The lack of existing transmission lines can limit the viability of a project depending on the project size. Take for example a large project site that has the available land to support 100 MW of solar generation. In an alternative scenario let's say that environmental restrictions only allowed 20 MW of solar generation to occur on the same project site. Assuming both installations would trigger the same interconnection upgrades, a 100 MW project would be more financially efficient as one can spread the interconnection upgrade costs over its entire project.

Ideally, solar power plants are located in areas where high levels of solar radiation can be taken advantage of to achieve higher capacity factors. Capacity factor is the ratio of actual power output to power output if the plant were generating full nameplate capacity for 100% of the year. The drawback to the ideal high solar radiation

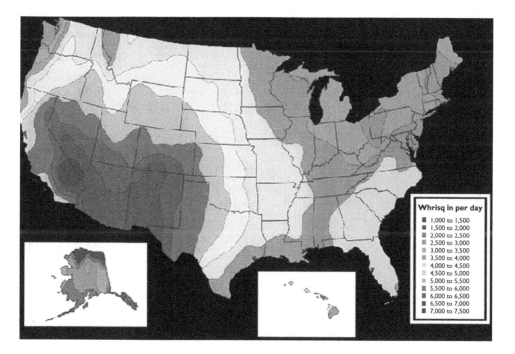

Figure 5.2 United States Solar Radiation Resource map (Courtesy of the NREL).

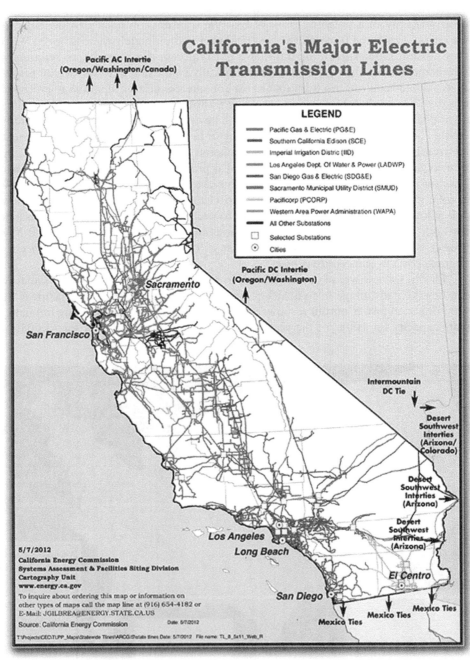

Figure 5.3 California transmission map courtesy of the California Energy Commission.

area is its geographic location. The remoteness of the desert southwest brings with it a lack of existing transmission infrastructure making it difficult to site or locate these solar plants. While there is an abundance of solar potential in the desert southwest, there are not enough transmission lines with available capacity to transmit the power cost-effectively to the energy end users. There are many committees today that are studying new transmission projects initiatives that would combat the lack of available transmission infrastructure. However, this problem is not likely to be resolved quickly as these committees are also facing the same issues as new solar projects such as difficult permitting arising from conservation of desert land and vital habitats.

5.1   LOCATING TRANSMISSION LINES

The first step in determining a project site's transmission viability is to determine if the site is in close proximity to transmission lines or substations; ideally transmission lines should be located near a potential project site and be large enough to support the planned generation. Then further investigation is required to identify how much transmission capacity is available in these nearby transmission lines. Finally, at a later phase determine how many upgrades, if any, would be required to support the expected plant output. Often these tasks can be addressed early in the development phase by transmission consultants and yield very useful results.

Several tools exist that can be used to locate transmission lines. One of the most accurate tools are transmission maps provided by transmission service operators. These maps are produced by groups who control and operate the electrical grid in a designated area like the California Independent System Operator (CAISO) in California, or Western Area Power Authority (WAPA) in parts of Arizona. CAISO is a non-profit public benefit corporation in charge of operating the majority of California's high voltage power grid. The costs of these maps are typically less than $200 and are available both in electronic (such as a GIS database) and print format. The maps often also include major roads and county boundaries that help to locate properties a developer may consider, while also listing the voltage and names of the high voltage transmission lines. Other options include third party maps and software like those provided by Platts, a provider of energy information and price assessments in the energy market. Groups like these often offer a very sophisticated GIS mapping tool that includes existing generators and transmission lines, as well as incorporating planned generation projects and transmission line projects. The accuracy of the information is reliable however, these are tools are usually targeted towards larger developers with many projects and potential sites in need of simultaneous review. The prices can be significantly above what we see for the local transmission maps from the utilities, but the capability of the software is more advanced and the information is more user friendly.

5.2   IDENTIFYING THE LINES AND NOMENCLATURE

Transmission lines are named by their start and end points with respect to substations and voltage levels. For instance, a 500 kilovolt (kV) power line extending from the

Mead Substation to the Perkins Substation would be technically referred to as the "Mead-Perkins 500 kV" line. Substation names include both a name, typically named after its location, and the voltage level. Hence, the Mead substation is referred to as the "Mead-500 kV" substation.

Once the name and location of a power line or substation is known and it is selected as a potential location to receive a project's power, the available capacity of the line must be determined. Executing a "queue check" or going through a third party transmission consultant are the two most common methods used to determine available capacity of a line. The queue check method is a very rudimentary technique that results in a "ballpark" answer of how much capacity is available just by checking how many projects are using or planned to use the line. This check can usually be conducted by individuals without a transmission specialty background, and would fulfill the needs of an expedited desktop analysis pre-screen for a project. It consists of going through the system operator's most recent queue and extracting the projects that have a planned interconnection point along the same transmission line or substation as the developer's planned project. What the developer wants to evaluate is if the planned projects already in the queue would exceed the line's thermal or conductor's kilo volt-ampere (KVA) rating. The more projects in the queue, the higher the likelihood that the power line, or area system will be very close to their design limitations and may require some upgrades. This queue check process roughly illustrates the potential available capacity of the line. In California,

Figure 5.4 120 kV (above) and 69 kV (below) three phase power line.

an online database system called the CAISO Transmission Queue exists where the amount of power, currently and planned to be, transmitted is provided and can be used to evaluate the transmission availability potential. The ideal locations for interconnection are the transmission lines or substations that have a high capacity and a low number of projects or power currently transmitted through them. Utilizing a third party transmission specialist is the simplest approach to gain a high level of confidence of the *actual* transmission capability. Alternatively, a transmission consultant can be used as a second phase to follow up on preliminary queue check filtering; a developer might not want to purchase engineering services and a formal transmission viability report for every site he comes across. After completing the internal prescreening analysis of the nearby transmission facilities and infrastructure, the transmission consultant can conduct the technical studies required for the available transmission capacities.

In order to best identify which transmission lines are adjacent to the project site, overlay the transmission map on top of the Assessor's Parcel Map. This allows one to accurately discern the one line targeted for interconnection. Sometimes if the prospective transmission line identified is not a verified line, pole numbers on the closest transmission line can be used to cross-reference. Transmission line structures look slightly different pertaining to different ratings of lines; knowledge of these types can help give an instant idea of the suitability of a site. Figure 5.4 represents two power lines with one of lower and one of higher voltage transmission lines, which can be visually distinguished because higher voltage lines have more insulators for added safety. Power lines are sorted into two sub-groups; "transmission lines" are those rated at 69 kV and above where "distribution lines" are those rated less than 69 kV.

Once a transmission resource is identified, and the project's requirements for capacity are specified, the next step is to complete an interconnection application and submit it to the interconnection or transmission entity. For most parts of California, an application would be submitted to CASIO. This application must include the application fee, description of the type of project proposed, exact location of interconnection, and the general electrical schematic of the facility.

5.3 HISTORIC INTERCONNECTION APPLICATION AND STUDY PROCESS

The interconnection study process in California is a great one to study as it has gone through the boom cycle of solar energy projects and had to adjust from an old model to a newer one. In the old California process but still in much of the United States there are two different application forms that differed depending on the project's nameplate capacity. For a project with generation capacity under 20 MW, a Small Generator Interconnection Process (SGIP) was the correct application form to complete and submit. For a project with generation capacity more than 20 MW, a Large Generator Interconnection Process (LGIP) was the application type and study process that the interconnection would have followed. In either study process, there were three primary study phases: feasibility study, system impact study (SIS), and facilities study. The interconnection feasibility study was the shortest study; 121 days

$10-20K Deposit	$10,K Deposit	$50K Deposit	$100K Deposit	$250K Deposit		Fund Upgrades Upfront
Application	Feasibility Study	System Impact Study	Facilities Study	LGIA	Restudies	Licensing & Construction
Time=0	Time=37	Time=158	Time=323	Time=534	Time=?	Time=624 (no restudies)

Figure 5.5 Typical 2008 LGIP timeline.

according to the LGIP CAISO process timeline from 2008 illustrated by Figure 5.5. This could have varied depending on the exact size of the project and the interconnecting entity (some irrigation districts handle this process internally and have their own procedures). The outcome of the feasibility study provided the applicant a top-level understanding of the ability of interconnection for the proposed generator to the existing transmission infrastructure. One-line diagrams and power load flow diagrams were utilized to complete this analysis. This process was directed by the interconnection entity.

If the interconnection customer decided to continue with the process, they entered into the SIS phase, which defined in detail the technical impacts of the generator on the system. Some required studies usually included, but were not limited to: transient stability, short circuit, heavy summer, autumn, or winter loads, and applicable generation options. The SIS phase studied the project in more depth than the feasibility study and took longer to be concluded. The results yielded a very clear perspective on the project in terms of interconnection impacts and effects. The study also took into account other existing electrical generators and those proposed in the transmission vicinity that would affect the ability to safely interconnect to the grid at the proposed commercial operation date.

The last study phase, and most costly, was the facilities study. This phase allowed the independent system operator to prepare the final studies meant to ensure that appropriate interconnection protections had been engineered to safeguard the electrical generator and the intermediary interconnection infrastructure. After this was completed and approved, the interconnection customer would have the necessary permission to interconnect the system as designed and specified through the SGIP or LGIP process.

The overall SGIP and LGIP process study timeframes varied depending on the amount of projects in the interconnection queue, and how many resources a transmission provider had available to perform the applicable studies. In of the 4th quarter of 2009, SGIP applications throughout the CAISO territory were projected to require 12 to 18 months of processing/study time, while LGIP applications were expected to take a more exhaustive 24 to 36 month processing time. While ultimately both these processes took longer in many cases (as there were new study requirements and many more applications than expected) understanding the processing times and backlog of applications provided project developers with valuable insight into the timeline required to a fully permit a project. Also, the difference between a 12 to 18 month and 24 to 36 month processing time generally swayed project developers to downsize the size of their proposed solar facility.

5.4 THE UPDATED INTERCONNECTION APPLICATION PROCESS

While the interconnection application will vary slightly for each entity, much of the information sought within the application is similar among different authorities. Depending on the voltage rating of the line that is planned for interconnection, the process may vary slightly. In 2005–2010 a significant amount of solar projects were requesting to be interconnected to the transmission system in California and the proposed processing times were taking nearly twice as long to complete. Hence, in late 2010 the interconnection stakeholders, convened to adjust the efficiency with which solar generation projects were studied. In the case of the utility PG&E, interconnecting to the grid at 60 kV or *below* will follow a "distribution interconnection" process and interconnecting to the grid *above* 60 kV will follow a "transmission interconnection" process. The processes vary due to safety, security, and oversight of the interconnection application and studies. Distribution interconnection applications are processed directly in cooperation with PG&E. Transmission interconnection applications are submitted to and managed by CAISO. In addition to the changing the process to better manages resources the utilities also have requested upfront payment of the entire study costs. Together this has helped to make the process more efficient and less speculative.

As the project developer will see, a generator interconnection request application can be broken down into three major parts: general application and applicant information, generating facility data, and transformer data. (PG&E, 2012) The Generating Facility Data will primarily come from the equipment supplier specifications, and some will have to be derived from interconnection engineering. As an example, the number of inverters will depend on the size of the inverter chosen based upon the capacity of the project with the inverter specifications coming directly from the supplier. It is very important to fully complete the application prior to submission. An incomplete application will not be given a queue position. An application completed inaccurately will be deemed incomplete. The applicant will be notified to submit the missing information in order to continue the application process.Taking PG&E as an example, the different interconnection study processes changed to align with one of three options: Fast Track, Independent Study, and Cluster Study.

5.4.1 Fast track

This process was intended for smaller systems seeking to interconnect 2 MW on a 12 kV distribution line or maybe 3 MW on a 21 kV distribution line. Up to 5 MW could go through this process on transmission level lines, but certain screens had to be passed and would have to be discussed directly with PG&E. The estimated time frame needed to complete all initial reviews and any needed supplemental review was 25 business days. As of October 2011, there was only a $500 application fee required.

5.4.2 Independent study

As its name suggests, this type of process will review and study the proposed project independently of other potential generators. Unlike the Fast Track process, the Independent Study process does not have power capacity restrictions on the distribution or transmission lines. It goes through more detailed transmission studies; a system impact study and facilities study are required bearing similarities to the old SGIP and LGIP processes. The study timeframe is significantly longer than a Fast Track process, occupying 120 days for distribution systems and 180 days for transmission systems. The application fee in an Independent Study is a tiered fee rather than a fixed fee like the Fast Track process. $50,000 in addition to $1,000/MW is the required application fee with a maximum cap of $250,000, the equivalent to the project having a net capacity of 200 MW. If the project is on the edge between going through a Fast Track process or an Independent Study, the developer would probably lean toward the Fast Track process to ensure spending less money on the application fee. As an independent generator, the cost for any upgrades or interconnection integration will fall upon the individual generator.

5.4.3   Cluster study

The prime difference between submitting into this Cluster Study process versus the Independent Study is that the necessary interconnection systems upgrade costs identified, if any, will be split evenly amongst the project in the cluster study group, which is composed of numerous projects. Similar to the Independent Study, the Cluster Study does *not* have a power capacity limit. The primary difference between Cluster and Independent is that a project that is within a Cluster Study goes through one annual study with a bundle of other projects that share the study costs. The application fee is exactly identical to the Independent Study as of October 2011. The two phases of the Cluster Study in total are estimated to require 330 calendar days for distribution or transmission interconnections.

5.5   LAND REQUIREMENTS FOR INTERCONNECTION

Planning and designing the solar field requires a substantial understanding of the interconnection permitting timeframe and likely interconnection upgrades required for a project. When submitting site plans to the local County Planning and Zoning Department, one may allocate space for interconnection requirements. These considerations include transmission easements for the generator tie line, and land for the construction of a substation if one is required. Sometimes, when submitting for these permits the exact details of interconnection are not known; in these cases it is best to be as detailed as possible with the county so that they are clear about the projects construction requirements and potential interest to the neighboring communities.

Figure 5.6 provides a good rule of thumb estimate regarding the different land requirements for various line taps and line loops into different voltage transmission lines. For reference, a line tap is defined as a new transmission line built to bisect an existing transmission line directly to allow for power to flow onto the existing line. A line loop is a transmission scenario where an existing line is cut and lengthened to allow for power to be added from a newly built bisecting line. The various tap and loop options do not require a significant amount of space but should be considered before detailed solar field layouts are finalized.

66 kV Tap: 90 ft by 110 ft

66 kV Loop: 110 ft by 150 ft

115 kV Tap: 100 ft by 120 ft

115 kV Loop: 120 ft by 180 ft

220 kV Loop: 290 ft by 480 ft

500 kV Loop: 420 ft by 780 ft

Figure 5.6 Typical interconnection infrastructure property requirements.

5.6 INTERCONNECTION COST ESTIMATES

Once the interconnection process has been initiated and the project specifics have been determined, both the developer and the interconnection authority will have a sharper perspective on the project, and what it will take to interconnect onto the grid. Requirements will be identified and at this point, the solar developer must incorporate these costs into their estimates. Completing the entire interconnection process will give complete clarity to the probable costs; however, a developer often needs estimates prior to the completion of these processes. In order to quote these costs there are "rule of thumb" estimates for the interconnection equipment. In its 2010 Solar Photovoltaic Request for Proposal, Arizona Public Service (APS) provided solar project bidders with estimated costs for common equipment to facilitate accurate estimations. Figure 5.7 displays the interconnection cost estimates that the utility APS provided has provided to assist energy bidders in preparing cost estimations. Finally, the network upgrade costs cannot be fully determined until the system impact study of an interconnection process has been completed.

For reference in 2009, Pacific Gas & Electric gave a presentation outlining interconnection cost in a California ISO (CAISO) stakeholder meeting. In their presentation they identified that the one-mile unit cost for a single circuit 115 kV transmission line would be $940,000, where if that same 115 kV transmission line were a double circuit, its cost would increase to at least $1.05 million. For a 230 kV transmission line, the one- mile unit cost for a single circuit is estimated at $1.1 million and at least $1.25 million for a double circuit type.

These costs were based upon the following principles:

- Flat land
- Rural area setting
- Normal soil
- Engineering and construction costs only
- Environmental, right of way acquisition, and permitting costs not included

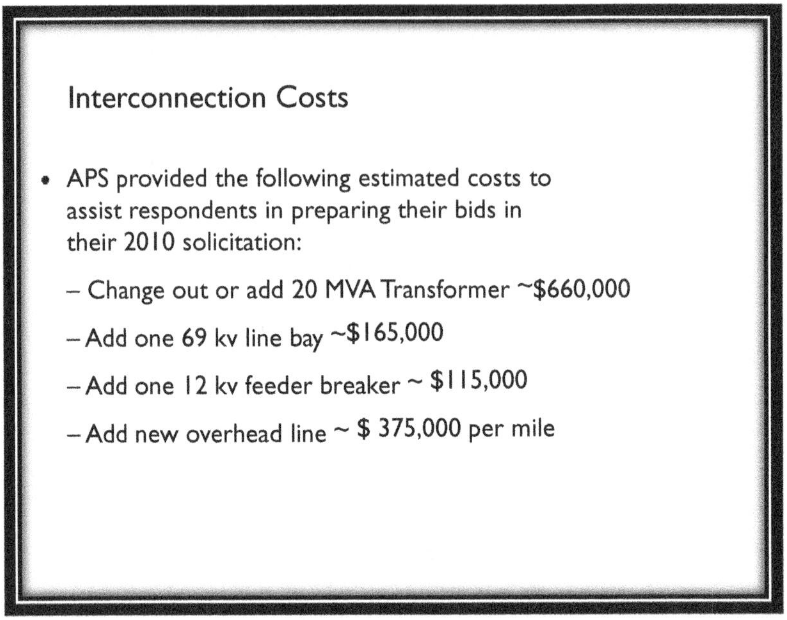

Interconnection Costs

- APS provided the following estimated costs to
  assist respondents in preparing their bids in
  their 2010 solicitation:

  – Change out or add 20 MVA Transformer ~$660,000

  – Add one 69 kv line bay ~$165,000

  – Add one 12 kv feeder breaker ~ $115,000

  – Add new overhead line ~ $ 375,000 per mile

Figure 5.7 APS Interconnection cost estimates for developers.

Table 5.1 Transmission cost factors for different types of terrain.

Characteristic	Transmission cost factor
Hilly	1.2
Mountainous	1.3
Forested	1.5
Suburban Population Density	1.2
Urban Population Density	1.5
Transmission Line Length Less Than 10 Miles	2.0
Transmission Line Length More Than 10 Miles	1.5

(Ng, 2009).

Other factors were assigned to different variables like terrain type, population densities, and length of the actual transmission line. These factors are identified in Table 5.1 but should only be used for budgetary purposes, and not used in place of actual engineering and transmission specialists.

Using these metrics, assuming there is a scenario that requires new construction of six miles of 230 kV single circuit transmission line to be built on hilly terrain in a suburban area, we would have the following cost estimate:

$940,000 (single circuit 230 kV transmission line) × 1.2 (hilly terrain) × 1.2 (suburban population density) × 2 (less than ten miles of transmission line length) = $2,707,200

5.7   APS TRANSMISSION CASE STUDY: TRANSMISSION

Arizona Public Service (APS) is the largest investor owned utility (IOU) in Arizona and has the largest electric customer base in the state. In its 2010 Photovoltaic Request for Proposal (RFP) period, APS identified an area where their constrained transmission areas existed within the state as shown in Figure 5.8. (APS, 2010) The constrained transmission area represents locations within Arizona where electrical generators, fossil based or renewable, would require significant transmission upgrades to deliver firm power to APS load centers. Hence, electricity produced in these areas would have a higher cost due to required transmission system upgrades.

APS serves a majority of the Phoenix vicinity and locations that are outside of their load center. Projects capable of delivering power to the load centers without going through a transmission constraint area have a higher level of priority, or value, to the APS procurement review team. Therefore, even though it is physically possible to wheel (wheeling is defined here as transmitting energy from a non-APS transmission provider's electrical systems to the APS electrical system), if interconnecting within the transmission constraint zone, APS would devalue the merits of the energy being provided. For projects within the constraint zone, APS is assigning a zero capacity value, equal to a 5 percent reduction in total energy output, to account for potential curtailments on the transmission system.

APS Constrained Transmission Areas

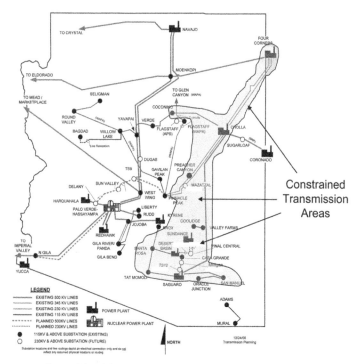

Figure 5.8 APS transmission constraint zone (APS, 2010).

From a solar developer point of view, it would be advantageous to propose a project that is situated in non-constrained transmission areas of APS to negate the wheeling charge and to provide the highest value of energy to APS without potential curtailments on their system. Even though wheeling is possible, the cost for wheeling would amount to approximately a 3 percent increase in the price of energy for this scenario. Wheeling costs are required to be paid, after commercial operation, for a determined dollar per megawatt cost every year of plant operation. Additionally, if APS devalues the energy being provided to them from a constrained area, the projects located outside of the constrained transmission area would possess higher viability and a better chance of being procured through a power purchase agreement (PPA). Strategically locating the project and understanding the intricacies surrounding load centers and potential energy off-take is extremely important to the early stages of permitting and development. The interconnection and transmission approach to a project can add as well as devalue the merits of a project depending on the utility's goal for procurement. Accurate estimates for the delivery of the power from a project is very important; neglecting to account for costs associated with wheeling can greatly change the price of power at any given delivery point.

5.8   TRANSMISSION CASE STUDY: NV ENERGY TRANSMISSION DELIVERY PREFERENCES AND IMPLICATIONS

Occasionally, a utility will specify locations where they'd prefer to see the delivery of renewable energy. This preference is based on transmission areas that are underutilized or have an excess amount of resource demand, and also those areas exhibiting potential for growth. In its 2010 RFP for renewable energy, NV Energy stated that its delivery preference points were: the Mead 230 kV, McCullough 230 kV, Harry Allen 500 kV, McCullough 500 kV, Crystal 500 kV, Midpoint 345 kV, Gonder 230 kV, and Hilltop 345 kV. As can be referenced from Figure 5.9, all of these preferred substation delivery locations are situated in the southern vicinity of Nevada. This preference is understandable as the majority of NV Energy customers are located in the Las Vegas area in the south of Nevada. Furthermore, many of these substations are strategically placed so generators could be located in Nevada, or nearby Arizona, and still provide deliveries to one of these points of interconnection without major transmission upgrades or added wheeling costs.

Connecting directly to one of the identified substations would be preferable to NV Energy and provide the highest level of viability from an interconnection point of view. If the project can't connect directly to the substation, being in the vicinity of a transmission line feeding directly to that substation would be the next best option. Fewer amounts of new transmission lines in need of construction naturally relates to lower risk. It follows that projects exhibiting a lower risk profile will gain a better project viability score by the energy off-taker. Major transmission upgrades are a potential risk to the viability due to the cost and permitting hurdles associated with building new transmission. If a project is relying solely on new transmission to be built, any delays to the timetable on which the interconnection upgrade is to occur could deal a fatal blow to the project. The fewer uncertainties there are to a transmission solution, the more willing a utility will be to procure the energy from a proposed project.

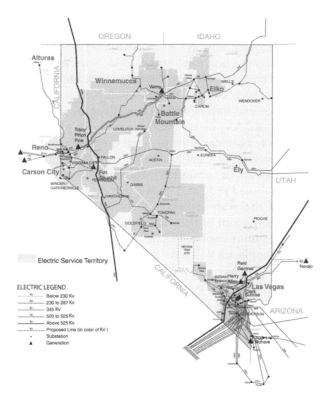

Figure 5.9 NV Energy transmission map.

5.9   CALIFORNIA INTERCONNECTION PROCESS – 5 MW FACILITY PLACED ONTO PACIFIC GAS & ELECTRIC NETWORK

Let us assume PG&E's Independent Study, updated interconnection application process, has been identified as the process required for this 5 MW project to interconnect onto the grid. This process deals with interconnections into PG&E's distribution network, identified as those facilities operating below 60 kV. This can be described as a five-step process consisting of the major parts of the interconnection.

The major phases are:

– Application Processing
– Technical Scoping Meeting
– Technical Studies
– Interconnection Agreement
– Project Implementation

5.9.1   Application processing

Application processing occurs once the developer has completed the interconnection application, and the application has been formally submitted to PG&E. During processing of the interconnection application, PG&E will evaluate the completeness of the application and make sure the application is accompanied by all of the application requirements. A successful application will have a fully completed application form that discusses and outlines all technical parameters of the project, and all equipment that is planned for use on the project. A site plan diagram must include the proposed project site boundary, the planned general layout of the project, and a proposed interconnection point (this can be pole numbers, general location of proposed interconnection point, or substation). The final technical requirement is a single line diagram of the proposed project. The one line diagram must portray the general electrical schematic of the proposed project, include safety equipment, and clearly demonstrate the inverter layout of the solar field. In the event that the application is submitted with missing information, PG&E will notify the applicant of the missing requirements, and the applicant will have time to respond to PG&E before the application is considered incomplete and rejected. Not until all information is completed and submitted will the applicant receive a queue position for the project.

5.9.2   The technical scoping meeting

The technical scoping meeting is conducted once the proposed project has submitted an application, the processing fee, and all of the technical requirements. The scoping meeting will provide the project's interconnection plans, which are commonly understood by PG&E, and define a point of interconnection and generator size. Lastly, during this meeting PG&E will provide technical system details, limitations, and insight on projects queued ahead, and will define the subsequent steps. When these steps are identified, good faith estimates will be delivered, and the cost will be further defined.

5.9.3   Technical studies

The technical studies commence after completion of the scoping meeting. Once the scoping meeting is concluded PG&E has a clear understanding of the project. All information required to model the project is gathered and the technical studies for the project will be initiated. All technical studies to be executed will clearly identify the reason for the study, the results that the study may yield, and a good faith estimate of the study cost. The studies will focus on capital improvements to PG&E's electric system to ensure the safety, reliability, and integrity of the grid. Additionally, studies will focus on the specific facilities required for interconnection and upgrades to the distribution systems triggered by the generator. The first technical study is most likely to focus on the impact of generation on PG&E's electric system; this study is often referred to as the system impact study.

5.9.4   Interconnection agreement

The interconnection agreement is issued at once all necessary technical studies are concluded and the project developer has chosen to continue with the interconnection process. The interconnection agreement will begin with a first draft of the agreement. Once this draft is issued to the applicant, the applicant and PG&E will have thirty calendar days to respond to the draft. Once the draft has been circulated, negotiated, and agreed upon, the applicant, PG&E, and the California Independent System Operator will execute it promptly.

5.9.5   Project implementation

The project implementation phase is the final step in the interconnection process, and this point is the ground-breaking phase of the procedure. Each party will be responsible for the items outlined in the interconnection agreement that was executed by all parties.

Chapter 6

Development: Energy off-take and power purchase agreements

Jesse Tippett

6.1 THE PPA AND ITS IMPORTANCE TO FINANCE AND VALUATION

Power purchase agreements (PPA) are one of the most instrumental and sought after milestones in a renewable energy project's development. This agreement makes or breaks the viability of a project; the financial difference between the value of a project with or without a PPA is paramount. A PPA is a binding contract between an electric utility and a project developer. In a solar PPA the utility commits to buy renewable electric power from the developer's solar generator over a long period of time, typically between 20–25 years. In the United States, these PPA contract prices are generally submitted on a competitive basis. They may either come in the form of a fixed price contract for the length of the PPA or with a yearly escalator over the contract duration. We will see that a PPA can have many varying negotiable terms and conditions, but its most important function is to define the cash flow for an energy project. It does not matter if the energy project is a renewable or conventional energy type, PPAs and other similar cash flow arrangements are critical to a project.

PPAs define the revenue that will be generated from a project and are the reasons why investors will evaluate the project for its investment potential. Without a PPA it is very difficult for the debt lender, equity participant, or tax equity provider to justify the value or the return potential for a given project. Usually investors will look at solar renewable energy investments the same way as any other investment; whether it is in a gas fired power plant, PV panel manufacturing facility, or investment into a gas station off the interstate. What investors care about is the initial upfront capital investment, the definitive revenue over a given period of time, and the level of risk involved to achieve the investment potential after the start of operation. In many ways for the solar project investment community, an executed PPA is just the starting point for determining the return potential for energy projects. There are a group of investors who will evaluate projects for procurement or acquisition without a definitive PPA. However the estimated cash flows should make conservative assumptions for the price to be received in a PPA (which also may never be awarded for any number of reasons).

6.2 BASIC STRUCTURE OF A PPA

The most typical arrangement for a renewable energy PPA is where two parties are involved, one buyer and one seller. Generally the seller forms some kind of limited liability company (LLC) which is owned by the project developer and the buyer is usually an electric utility company. The LLC owns all of the assets of the project including: land site control, permits, solar generating equipment, project related engineering documentation, and the PPA contract. Due to renewable portfolio standards, electric utilities enter into PPA contracts for solar because states are regulating their utilities to procure a certain percentage of their energy generation through renewable sources; solar has been one of the most popular procured renewable energy types in recent years. In this agreement, the seller agrees to sell electricity in the form of kilowatt hours (kWh), which usually comes bundled with renewable energy credits (REC). The bundled energy attributes associated with the electrical energy can come in the form of green credits, such as carbon credits, or RECs. Even the energy delivered has additional benefit to the utility since there are certain parts of the day where energy is more expensive for the utility to buy and resell it to its customer base. The energy is priced on a kWh unit basis and in some PPAs, the seller is paid on a sliding scale depending on the time of day when electricity is delivered to the grid; this is known as the time of delivery payment structure and is common in the southwestern United States, especially in California. In general, the energy buyer pays a higher rate during their seasonal, daily, or hourly peak demand time periods.

The basic representations of the seller are a type and quantity of energy that will be delivered on an annual basis. The energy promised shall be from a certain location, at a specified point of interconnection, for certain period of time, and at a predetermined price. On the buyer side, the representations are that the energy will be purchased under a certain set of conditions for the lifetime of the PPA, at the given point of delivery for the agreed upon price. Both sides represent that they are in good corporate standing, creditworthy, and have the capacity to fulfill their respective obligations under the negotiated PPA conditions.

6.3 KEY TERMS OF THE PPA

Most PPAs today are considered "boilerplate," meaning contracts that have evolved over the years with a number of standard terms and conditions. However, there are a certain number of key items that differ from contract to contract and are very important to the developer, potential investors, and even the utility or buyer of energy.

The six most important terms of a PPA are:

1) The contract electricity price
2) The annual escalator
3) The term of the contract
4) The location delivery point of the energy and generation facility
5) The description or definition of the type of generating facility
6) Contract off-ramps (curtailment)

These six terms differentiate one project from another, and may also spell the difference between success and failure of an energy project in the development stage. The contract electricity price, annual escalator, and term of the contract are long-term financial commitments for the buyer. While it is beneficial to gain price stability for a long period of time, the buyer wants to ensure they are not getting into a deal that puts them in a bad financial situation for a lifetime. What is of interest to utilities is that solar technology is rapidly improving in efficiency and cost therefore, executing PPAs strategically allows for potentially beneficial electricity prices, which may lead to more buyer friendly escalation rates and contract lengths.

The location and delivery point of the solar generating project is extremely important as the utility has to schedule how energy is getting onto its grid and must act accordingly to carry that energy to population zones and load centers. The description of the facility type is important because different solar technologies have vastly different energy generation profiles. If a utility was expecting energy delivery from a concentrating solar power plant which provides fairly stable energy profiles, they would want to define that in the contract. The buyer wants to avoid a situation where the project owner switches the technology type, which may result in undesirable energy profiles. Finally, contract off-ramps (such as curtailment clauses) which dictate where an off-taker has the ability to cut or reduce the payments agreed upon within the PPA can cause financeability issues.

6.4   FINANCING CONSIDERATIONS OF THE PPA   .

As discussed earlier, a PPA is the keystone to an energy project's success. It is paramount that the PPA is financeable; this is easily stated however, it can be complicated and difficult to achieve a PPA that embodies this single idea. Perhaps one in ten energy projects, not only renewable energy projects, ever get built. This is often related to fatal flaws in the financeability of key agreements, with emphasis on the PPA.

When a potential stakeholder reviews a PPA for financeability, one of the first questions asked is "what is the credit rating of the off taker?" Most large North American utilities pass this test with flying colors; however, this question becomes more important when looking at projects involving a small, private, or inexperienced off-taker. Subsequent considerations are: determining if the term of the contract is long enough and matches the term of other key agreements such as the site control agreement, and determining if the price and projected generation will produce a return that can support the capital expenditure of the project.

Developers should be careful to ensure that subtle fatal financeability flaws are not present in their PPA contracts. In general, these financeability flaws often revolve around situations where the buyer does not need to or cannot perform their key obligations, such as accepting and paying for the generation of an energy project. Key areas that often affect this financeability are curtailment, whereby the buyer is excused from accepting or paying for energy from a project usually due to the unavailability of the transmission network or other defined excuses. The developer should also be aware of the subtle fatal flaw of ambiguous or uncertain regulation compliance payment terms. These terms are often defined as payments that the project selling energy may have to pay in the future, should regulations or laws be enforced in a different way. If either of these types of subtle fatal flaws cannot be completely dismissed, it is often a good approach to limit or cap their quantities and impacts. Hence, a potential remedy would be to limit compliance payment to an annual fixed amount, or curtailment could be limited to a certain number of hours per year. Developers may work to remove these concepts entirely.

6.5   PPA VARIABLES GREATLY AFFECTING THE CASH FLOW AND VALUE OF THE PROJECT

While the entire PPA is important, certain variables greatly affect the cash flow of a project and its potential for a positive, investment grade return. These variables include the price, the price escalator, the term of delivery, and the payment structure (time of delivery, etc). Furthermore, the varying combination of these key items can produce vastly different cash flows for seemingly similar figures. For instance, by utilizing the NREL System Advisor Model (SAM) software, a simple analysis was performed for a standard size project (20 MWdc with a 20 year PPA term) modeling a 0 percent escalator that concluded a price of $174.10/MWh was needed to achieve a targeted internal rate of return (IRR). However, when a 2 percent yearly escalator is sought, a price of $133.60/MWh is sufficient for that same IRR. With this slight change, one can see how all factors of a PPA price must be weighed *together* to determine the real value. This nearly 30 percent swing in the initial PPA price will most likely mean the difference in winning or losing a PPA. Were this escalator not included at the lower price, that scenario would most likely not allow for a financeable project return. The developer should be advised to consider all of these factors together when proposing a PPA, valuating economic potential, and developing a project in order to achieve a desired monetary return on investment.

As the American solar market has matured, utilities have awarded contracts to solar projects with more competitive pricing for its source of renewable energy. The reason for the lower pricing has been a function of decreasing solar equipment costs, competitive bid processes, and a highly competitive market gaining more entrants as the years have passed. This makes renewable energy projects economics more and more challenging, and requires developers to be much more accurate with their assumptions and cost estimates. As of late 2011, the price of solar renewable energy is coming down; however, it is not yet cheaper than generation from traditional sources, such as natural gas or coal, with all market characteristics being equal. From a utility and public policy perspective, the reader may already be able to surmise that by adding an escalator or increasing the contract term, energy prices could be achieved that are lower than prices without an escalator or those accompanied by a shorter term contract. The base resources for renewable energy are natural forces such as the sun, the wind, and the availability of waves. These are arguably much less variable than commodity prices related to natural gas, petroleum, and coal. Requesting or including an escalator or longer term in a power purchase agreement to coincide with normal economic terms such as inflation, could help present renewable energy prices on par with traditional energy sources. However, at present the utilities continue to pursue the lowest cost energy with lowest levelized cost to meet its RPS obligations, and fail to value the inherent predictability of the cost of solar energy.

The term length of a PPA is not only important to achieve a project return, but it is also especially important for the debt financing component of a project. Hence,

PPA terms in conjunction with the energy price should be long enough and priced accurately to yield a sufficient project return *even if* unexpected circumstances such as facility energy underperformance or force majeure events arise. In the highly competitive world of PPA bidding there is often not a lot of room for large contingencies so prospective developers are advised to discuss and analyze these scenarios within their teams. In general, a longer term of a contract is often a good way to cover for multiple contingencies, especially related to the energy price, while not adversely affecting the PPA from the utility side.

6.6   BUYER AND SELLER KEY COMMITMENTS UNDER A PPA

For a developer, two key items stand out as the most potentially risky and difficult to perform under a PPA. These are the guaranteed construction and commercial operation dates, and the guaranteed annual energy output. Regarding the former, often in the development process, a project has not completed its permitting or interconnection process when it is awarded a PPA. In many cases, it takes at least six to nine months to submit the PPA proposal, get shortlisted by the utility, negotiate PPA terms, and finally provide financial security deposits along with the full execution of the PPA. Within the negotiated PPA are definitive dates stating when the guaranteed start of construction is to occur and when the guaranteed commercial operation is to commence. There are delay clauses that would allow those guaranteed dates to extend into the future. Any delay in gaining permits and approvals to commence with construction would cause the expected commercial operation date to slip. Missing the commercial operation date often has a consequence of liquidated damage payments to the off-taker utility by the project company. In the worst-case scenario, an extreme delay could lead to contract termination if guaranteed milestone dates are missed. Extensions to the guaranteed milestone dates should only be used as a necessary cushion bearing in mind that coming too close to the point where PPA contract termination is a possibility that might lead to a lack of investor interest.

Another risky item to perform under a PPA is guaranteed annual energy output. To maximize the solar field's potential expected generation the developer must work to gain a technical understanding prior to the PPA proposal. The more electrical generation that a solar project can output at a given fixed cost, the more flexibility the developer has with providing a more attractive PPA electricity price. This has to occur while balancing the pressure to present the most energy possible to utilities, which leads to a larger creation of RECs to meet the state's RPS, and delivering energy during periods of the day that gives them the most capacity value. Typically, if the actual generation is less than the guaranteed annual energy prediction stated in the PPA the result would be a liquidated damage or contract termination if actions are not taken to rectify the problem. To cover the utilities risk for both aforementioned items, the utility will normally require a development and operating term security under the PPA contract. Development securities often range in the amount of twenty to thirty dollars per kilowatt of generation capacity. Term securities are often a one-time posting of 50 to 100 percent of revenues of one average contract year.

6.7   OBTAINING A PPA

In general, the PPA process can be described chronologically to begin with the opening of a request for offer (RFO) or request for proposal (RFP). For major California utilities, this has occurred at least on an annual basis due to the high need to meet aggressive RFP targets. For smaller utilities with smaller RPS targets, this may only occur on an "as need" basis depending on the development on previously contracted projects. In projects without a power contract, a prospective developer would review the utility's RFP to determine if an opportunity existed to propose its project. Usually these RFPs are made somewhat transparent prior to the official release allowing developers to provide the off-taker with a renewable resource that best fits its energy needs. Next, the developer would generate a response complying with all the requirements of the off-taker's RFP. An RFP requires various project related information and typically involves: the definition of the project, project location and technology type, energy delivery point, annual generation, hourly generation profile, summary of developer and stakeholder experience, key attributes to project viability, the price, term, and price escalator. Should the utility, after its internal review of possibly hundreds of similar proposals, determine that the project met their requirements, a notice of success or short listing would be issued to a limited number of applicants. At this point the developer is invited to execute a form of PPA and post a development security. Acquiring or achieving a PPA is often the crux of energy project development. This is not an easy task; it involves a multitude of disciplines and takes place in one of the most competitive business environments due to the high and stable return potential.

Development or PPA signing securities, as discussed earlier, are often in the twenty to thirty dollars per kilowatt range (about $400,000 to $900,000 for a standard 20 MW photovoltaic project). One can see how what may start out with some developers as "wild catting" quickly turns into a balancing act of strict contractual requirements and large amounts of upfront cash. While utilities often do not allow large-scale changes to a form PPA (as these are often a PPA pre-approved by regulatory agencies), the period between short listing and signing is the correct time to request any changes a developer could possibly require. Often these changes will relate to items regarding PPA contract financeability, such as lender consent terms and technical clarifications.

After negotiation and execution of the PPA, it is up to the developer to finish the development of a project, to obtain construction financing, complete the construction, and begin delivering energy requisite with the obligations of the PPA. As these PPAs are often obtained at the early stage of a project's development, this is much easier said than done.

6.8   UTILITY CONCERNS IN PPAS

Utilities should and often do have several key concerns when reviewing PPA proposals. However, experienced developers would indicate that to date utilities review and award PPA proposals based solely on how low the energy price is. This has resulted in the award of many contracts at low prices yet the public is seeing few projects that actually come to fruition. A host of historical examples are available to support this such as the PPAs awarded to Stirling Energy Systems and Optisolar projects. These projects totaled over 1000 MW and have changed ownership and technologies but have yet to be constructed. With that said, utilities are becoming more sophisticated in their approach to renewable energy by applying much of what they have learned from their years in traditional generation. Hence, as of late utilities have placed higher value on the level of project viability including factors relating to developer experience. While this added level of evaluation is primarily in the form of subjective score weighting and judgments, utilities have also increased financial requirements for seller performance in the proposal, development, and operation portions of PPA bidding typically in the form of security payments.

It is yet to be seen if the utilities revamped review process will affect the construction of a higher volume of projects. The current market landscape provides a plethora of project options for utilities to choose from compared to the time of when Optisolar and Stirling Energy Systems submitted their PPA bids. The market is challenging, and it should be noted that developers getting started in renewable energy often see winning a PPA as a lifeline to the next level of investment. It is this common contemplation that has created a highly competitive market where bids are submitted in an almost desperate fashion. The result of this is prices that are set extremely low, even artificially low, and these "deals" are hard for the utilities to pass up. The prices are often so low that not even the most efficient companies, utilizing the most state of the art technologies and financing means, can generate above water financials with a market level winning PPA price.

6.9   KEYS TO PPA AWARD

In general, the key to winning a PPA is presenting a project that displays a realistic and viable path to completing development on time, on budget, that delivers energy at a competitive price, and with a technology that can confidently work throughout the term of the contract. The soundest way to present the developer is by highlighting their experience; however, as solar power is a maturing industry, often the developer experience section of a proposal consists of key team member biographies rather than company history. A real success story of winning PPAs relates to the project development company Nextlight, which does not own any intellectual technological property. Just by going to their website and reading the thirteen executive member biographies, its feels as though you're reading about the "Solar League MVPs." Experienced team members can and have helped firms achieve success in this market. This is one of the key reasons the Nextlight team was able to propose, win, and ultimately sell several hundreds of megawatts of PPAs within such a short period of just a few years.

6.10   FINDING THE BEST UTILITY FOR A PPA

When considering the renewable energy market as a whole, a developer might ponder which off-takers give the best power purchase agreements. There are two basic requirements, credit rating of the off-taker and the ability to negotiate a financeable PPA. Geographically there are no clear, hands down winners. In a broad sense, inexperienced rural utilities and entities often create some of the most developer friendly forms of PPAs. However, as they become educated, and they quickly do, these agreements often evolve into the larger and bureaucratic style contracts that are found with large investor owned utilities (IOUs).

6.11   PROJECT MATURITY AND PPA AWARD

As the reader has already surmised, the renewable energy development process is complicated, and you may ask "When is it appropriate to target a PPA?" Two diametrically opposed points of views exist around the answer to this question. One perspective proposes that a PPA should be achieved at very early stages of project development potentially just after achieving site control, with only minimal submission of permits (if any), and with the minimal amount of engineering needed to produce hourly solar power generation results. The opposing perspective suggests waiting until development is nearly complete and there is a real light at the end of the tunnel. The latter approach is certainly the safer approach; however, even a large resourceful project development firm receives pressure exerted by the market to win PPAs since significant development capital is being spent in the meantime. At a minimum before proposing a PPA, a developer should have site control, clarity of the costs associated with the interconnection of the project, and a definitive timeline as to how the project will reach their proposed commercial operation date (COD). If the developer has

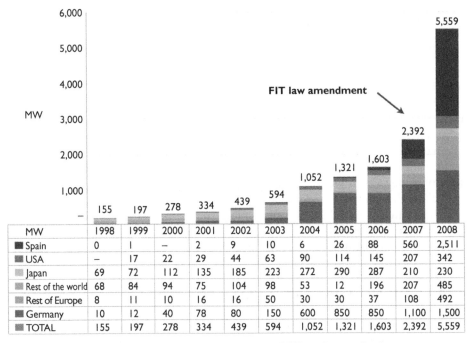

MW	1998	1999	2000	2001	2002	2003	2004	2005	2006	2007	2008
■ Spain	0	1	–	2	9	10	6	26	88	560	2,511
■ USA	–	17	22	29	44	63	90	114	145	207	342
■ Japan	69	72	112	135	185	223	272	290	287	210	230
■ Rest of the world	68	84	94	75	104	98	53	12	196	207	485
■ Rest of Europe	8	11	10	16	16	50	30	30	37	108	492
■ Germany	10	12	40	78	80	150	600	850	850	1,100	1,500
■ TOTAL	155	197	278	334	439	594	1,052	1,321	1,603	2,392	5,559

Historial development of the global annual PV market per Region

Figure 6.1 Global PV Development (The Efficiency of Feed-in Tariffs in Germany and Spain, Amin Zayani, Feb, 2010).

sufficient experience to generate this information reliably, then arguably these would be the only prerequisites to bidding a PPA.

The PPA structure presents interesting challenges. Amidst these challenges, a developer is often locked into generating energy at a specific geographic location by a specific date. Alternatively, under an arrangement such as the feed-in-tariff (FIT) a developer can enter into a standard offer contract, which has a predetermined energy price and terms. These types of contracts were very common in Germany and Spain. In a FIT power contract, developers do not feel many of the market pressures as they have the ability to change sites and also COD within a larger window. While there were many shortfalls to the FIT, we have noticed in the past how effective these structures were for implemented solar energy. For instance, in 2008 alone Spain installed more than 2.5 gigawatts of solar energy generating systems (Couture, 2011).

6.12   THE BROKEN U.S. PPA MARKET AND KEYS TO IMPROVING IT

The PPA structure requires that both the utility and the developer assume ambiguous risks at a relatively early stage of the development process. While this is nothing new in development or business, the generally low PPA prices created by the especially competitive American market have been the key stumbling block towards wide-scale installation of renewable energy projects. One may say when the price is low enough the market will do the rest, but it is known that there are a host of reasons and intangible benefits renewable energy has including producing energy that is clean and contains no carbon. The real key reason that there has not been a significant uptake in renewable energy in the United States, one of the richest and most developed countries in the world, is that these intangible benefits have been left on the table. The avoided cost inherent in non-polluting energy has not been effectively priced into the value of electricity derived from renewable sources.

The PPA structure is unlikely to be done away with in the United States for a variety of reasons; however, liquidity and demand for green credits, such as tradable RECs or carbon offset credits, could have a great long term effect on sustainably supporting the development of renewable energy projects.

Chapter 7

Development: Renewable energy credits

Katherine Ryzhaya Poster

Environmental commodity trading is a market-based approach aimed at controlling pollution at the lowest possible cost by providing economic incentives for achieving emission reductions and penalizing non-compliance.

7.1 RENEWABLE ENERGY CREDITS

Renewable Energy Credits (REC) represents environmental and social benefits of renewable energy generation. RECs serve as certificates of proof that one unit of renewable energy has been generated. Most commonly, the units are measured in MWh, where 1 REC = 1 megawatt hour (MWh) of renewable generation. RECs are also the accounting tool used to prove that electricity sellers have complied with the Renewable Portfolio Standard (RPS).

Industry has been aided by regulations at the state level mandating use of renewable energy and by the push among corporations to offset their environmental impact by "greening" their power consumption. As with most robust, functioning markets, financial instruments develop to provide transactional ease, hedge risk, and reduce the cost of compliance. In the case of renewable energy markets, nearly all active jurisdictions have adopted the REC commodity in an effort to accurately measure, quantify and verify compliance with renewable goals. In this chapter we will define the mechanics supporting the trade and valuation of these commodities and show how they serve to quantify the often 'higher' value of "green" generation.

7.1.1 Tradable instruments

The process of generating renewable energy creates two separate products, electricity and RECs. The two products may be sold together as a "bundle" or separately. Unlike electricity, RECs can be purchased and sold without the constraints of the physical power market. Purchased RECs are subsequently "retired" in demonstration of compliance with environmental mandates or in conjunction with corporate commitments.

Two separate buyers may be purchasing the electricity and REC components from the same renewable generator as shown in Figure 7.1. The buyer of the electricity piece has in essence purchased non-renewable "brown" power. This buyer cannot make environmental claims against the power since the REC was not included in the purchase.

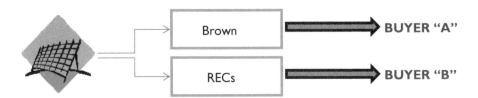

Figure 7.1 Products of renewable energy brown power and RECs.

While some compliance jurisdictions favor "bundled" (electricity plus REC) transactions, it is more efficient to allow RECs to be treated as separate commodities. In many instances, the physical location of the renewable generator is not within proximity to its buyers. By separating the environmental attributes of renewable generation from the underlying electricity, buyers can support the growth of renewable energy regardless of physical location. A greater pool of resources also contributes to price competition and allows the generator to lock in the best value for its products.

RECs are purchased and sold primarily in bilateral, over-the-counter (OTC) transactions, through voice brokers or directly from counterparties.[1] Retail marketers serve the needs of individuals and corporations.

The volume of RECs traded in the marketplace at any given time may depend on a number of factors, including:

- Whether multiple vintage year or futures transactions qualify for compliance
- Whether RECs may be banked (i.e. carried over) from one compliance period to the next
- Whether restrictions are put in place with regard to geographic and operational characteristics of participating generators
- Pricing expectations
- Timing of the next compliance assessment period

In the U.S. the absence of a federal renewable energy target leaves program oversight to each participating jurisdiction. Generally, for investor-owned utilities such as Duke Energy in the Southeast or Pacific Gas and Electric Company in the West, the local Public Utilities Commission (PUC) would define program parameters, oversee all contracts and marketplace transactions, and regulate any impact to customer rates.[2] Public utilities generally fall outside of PUC jurisdiction and are governed by their local boards.

The eligibility of REC generating facilities is defined either by law or by the appropriate state-level regulator. Designated agencies track generation from sources that register in the system and facilitate a platform for transfers or exchange. REC certificates are used to verify compliance with regulatory requirements and in voluntary market programs.

---

1 The Intercontinental Exchange ("ICE") is the only trading platform that currently lists REC products: New Jersey Class 1RECs, Massachusetts Class 1 RECs, and Connecticut Class 1RECs.
2 In California, the Public Utilities Commission and the Energy Commission collaboratively implement the RPS.

7.1.2 Regional markets (U.S.)

Commoditization is a feature of many mature renewable energy markets. In the U.S. steady volumes of transactions, relative price transparency, strong and reliable regulatory signals, and tenured participants categorize eastern markets. In these jurisdictions, compliance with RPS mandates is achieved almost exclusively through RECs. These well-balanced trading systems are designed to reflect the aforementioned realities of the transmission grid; the best locations for power generation are often far removed from populated urban consumption centers, making the physical delivery of generated electrons either cost prohibitive or entirely impossible. By unbundling the REC from the underlying power, the buyers of green attributes in the Northeast are contributing to "greening" up the power mix by enabling a renewable resource to displace a higher polluting generator.

In the West, California's renewable energy program is one of the most ambitious mandates in the country. The California RPS requires utilities to obtain 33 percent of their retail sales from renewable energy sources by 2020. This mandate translates to nearly 75 terawatt-hours (TWh) of renewable energy generation.[3]

Until recently, California did not have a defined REC trading component in its RPS arsenal. All transactions needed to be fully bundled in order to qualify for the RPS. However, high levels of demand far surpassed the amount of resources connected to the California grid resulting in California utilities forced execution of complex, swap-like arrangements in order to reach remote supply and attain compliance. Some such structures still isolated the REC component from the underlying energy. Figure 7.2 depicts one such transaction, where the utility purchased the bundled RPS product and simultaneously disposed of the electricity component by selling it back to the generator. The utility then matched the RECs with system energy sourced under a second contract. It took several years of stakeholder opposition to rename this transaction as "REC-only."

Given the sheer number of environmental and social interests represented in California's renewable energy market, trading in the western U.S. will likely never be as liquid or commercially-minded as in the northeast. Drivers outside of basic economics will need to be considered when sorting supply. The plethora of rules dealing with origin, vintage, transfer, and tracking will pose further liquidity hurdles, as well as the various product categories and segment caps. The California RPS Compliance Instruments literature in Section 7.1.3 outlines basic guidelines for California's renewable eligibility classes.

7.1.3 Commercial and regulatory risks

As in any financial market, REC counterparties are faced with a host of risks in the course of business. In addition to credit (counterparty default) risk, which is expected in bilateral transactions with no central clearinghouse, buyers and sellers are exposed to significant commercial and regulatory risks. Some of these risks are discussed as follows.

---

3 California Public Utilities Commission (2009) *33% RPS Implementation Analysis Preliminary Results.* [Online] Available from: www.cpuc.ca.gov/NR/rdonlyres/1865C207-FEB5-43CF-99EB-A212B78467F6/0/33PercentRPSImplementationAnalysisInterimReport.pdf

---

California RPS Compliance Instruments

In accordance with Senate Bill 1X-2 ("SB2"), California utilities are required to procure an average of 20% of their load from renewable sources for the period of January 1, 2011 to December 31, 2013; 25% by December 31, 2016, and 33% by 2020. SB2 also created three distinct product categories within each of the three compliance periods.

Volumetric caps apply to procurement in each of the three categories, below.

Category 1: In-State/Dynamically-Scheduled: Renewable resources directly connected to a California balancing authority or provided in real time without substitution from another energy source.

Category 2: Firmed/Shaped Products: Energy not connected or delivered in real time yet still delivering electricity.

Category 3: RECs: Unbundled/unattached renewable energy credits.

Resource allocation to each bucket differs with each compliance period. In an effort to capture/internalize environmental and social benefits of the RPS program, *procurement allocations to out-of-state resources diminish over time.*

(California Legislature, 2011)

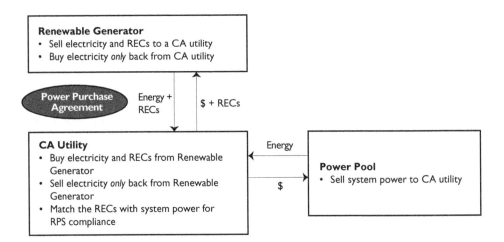

Figure 7.2 Example flow of energy RECs and $ in a renewable energy project.

- *Market Illiquidity:* The lack of comprehensive regulations and inconsistency in product definitions preclude a robust market with active participants and price transparency. The net result of regional fragmentation is constrained demand, uneducated supply, and the lack of new resource construction. Improving liquidity includes standardizing product definitions, connecting markets, and normalizing rules (tracking, ownership, etc.) to apply equally to all participants.

- *Pricing Ambiguity:* In transparent, tenured markets, the price of RECs is easily discoverable. Other newer, more complex markets do not provide immediate transparency into pricing and volumes of either supply or demand. Instead, developers are encouraged to submit "best bids" which are subsequently evaluated by utilities against all offers received in a particular solicitation. Winning bids are not made public in order to preserve utility ratepayer value. Developers whose bids were not selected do not know with any accuracy what price would have made their bids competitive.

- *Production Requirements:* Renewable energy projects with intermittent generation profiles (solar, wind, hydro) often rely on REC revenue to justify investment and operations. Poor performance periods will yield less revenue than projected. The issue could be further exacerbated under "firm" delivery contracts, where a guaranteed REC volume is conveyed to the buyer. Not only will the REC seller lose performance-related revenues, he may also be subject to shortfall penalties per the REC contract. (Unlike a "unit-contingent" delivery contract, which only conveys those attributes actually generated by the facility, firm delivery commitments obligate sellers to deliver guaranteed amounts of RECs. Utilities subject to RPS compliance are reluctant to sign unit-contingent deals due to the threat of volumetric shortfall in a given compliance period.)

- *Counterparty Diversity:* Illiquid markets are often defined by constrained participation and counterparty concentration. From the utility perspective, a small supplier pool would preclude sufficient risk diversification. For suppliers, the lack of buyer interest (due to massive competition for the same counterparties) causes price deterioration beyond the point of financeability. Project actualization is challenged under both scenarios, whether in response to pricing pressures or market saturation.

- *Product Eligibility:* All "compliance-grade" REC transactions are executed in accordance with the relevant RPS. While legislation defines product eligibility; program implementation is left to the local governing agencies. To date, regulatory uncertainty stemming from program infancy, lack of coordination, and mid-game rule changes has imposed substantial risk on transacting parties. Utilities have started looking to sellers to guarantee RPS eligibility of their products for the duration of the contract, often up to 20 years. Under this construct, any subsequent change in regulation becomes a seller liability and contract sanctity is challenged when commercial transactions are at risk of post-execution modifications.

- *Regulatory Approval:* This risk, associated with utility-scale rather than smaller commercial transactions, is the potential for deal rejection by the regulator in charge of rate-basing *after* the purchaser has signed the contract. In this case, the regulator prevents the commercial transaction from taking effect and being charged to utility customers. The purchasing utility often hedges this risk through a "no fault out" clause. The REC seller, however, stands to lose the contract and substantial amounts of time, effort, and investment.

- *Inter-Market Coordination:* The RPS program of any given region does not work in isolation. Other interests, such as technology-specific carve-outs, resource adequacy requirements, and competing climate concerns often complicate the administration of a given program. In California, where the State's Assembly Bill 32 has codified a carbon emissions reduction requirement, questions arise as to how RECs will be treated within the cap-and-trade construct. The central question is whether a REC that was traded separate from the underlying electricity will carry a renewable (i.e. zero) emissions characteristic or a system average emissions rate for carbon compliance purposes. The issue is further complicated by the fact that while RPS and carbon programs exist to achieve similar environmental and social goals, they are separately managed and run. When too many separate government branches are tasked with administering rules and measuring compliance of similar programs, one can expect a certain level of ambiguity and misalignment with commercial realities.

Several years of REC trading data provides insight into certain market dynamics. In New Jersey, oversupply of solar RECs (SRECs) has led to a dramatic drop in price in the spot market. In the early 2000s, the NJ solar market was a lucrative place to be; the combination of state and federal incentives presented an attractive package for homeowners who could suddenly afford solar's up-front price tag. Coupled with a trading system where compliance-bound utilities purchased high-priced SRECs, capacity more than doubled and the market became over-saturated. The net result was a dramatic drop in price (nearly 70 percent) and a stranded pipeline of projects.

Likewise, Pennsylvania's formerly thriving solar industry is suffering. With installations outpacing the RPS mandate, SREC prices are falling faster than anticipated. (Pennsylvania's price/participation dynamics are, however, subject to change. Energy companies PPL and PECO represent over 50 percent of the electricity in the market and both have been exempt until 2010 for PPL and 2011 for PECO. Their participation in SREC trading is likely to have a positive impact on price.)

The negative effects of market saturation are plentiful around the world. While in some jurisdictions it is pricing that suffers, in others entire industries are halted due to oversupply. Germany, a known leader and supporter of solar energy, has cut its subsidy in the 1st quarter of 2012 for new installations by 30 percent. The government had explained its decision as a means of slowing the rapid growth of the sector, saying the industry has been allowed to grow too fast and had been too heavily subsidized. (Connolly, 2012) This change in the level of financial support had almost instantaneous effects (German companies Q-Cells, Solon, Solar Millennium and Solarhybrid have all filed for insolvency) and will undoubtedly have long-lasting ramifications for the solar industry in general (Reuters, 2012).

7.2 CARBON MARKETS AND CAP-AND-TRADE PROGRAMS

7.2.1 Tradable instruments

The cap-and-trade system depicted in Figure 7.3, establishes a cap on annual emissions, identifies those entities whose emissions will be regulated, and provides a means for the transfer of allowances among parties. The rules of trading are typically set by an exchange or convention in the OTC market. Markets are also overseen by regulators such as the U.S. Commodity Futures Trading Commission (CFTC).

Compliance is achieved using a combination of direct emission reductions, as well as the purchase of two compliance instruments, *allowances* and *offsets*.

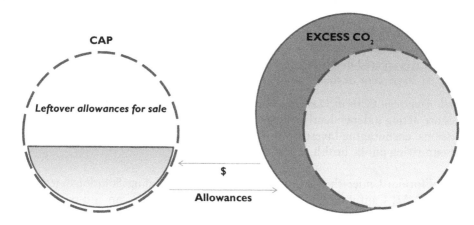

Figure 7.3 Mechanics of carbon trading, over compliance with regulations produces surplus allowances that can be sold.

- *Allowances:* Regulators establish an industry cap on emissions and allocate, or auction, allowances to emitters and other market participants. These allowances, the equivalent of one ton of CO_2, are the actual compliance instruments that emitters use to meet their obligations under the emissions cap. At the end of a compliance period, emitters will submit to the appropriate regulator allowances for each ton of CO_2 they emitted during the period. The amount of allowances is reduced over time.
- *Offsets:* Cap-and-trade programs often include a compliance-offset component. Offset credits are greenhouse gas (GHG) emission reductions or sequestered carbon that meet regulatory criteria and may be used by an entity to meet up to some defined percent of its compliance obligation under the cap-and-trade program.[4] Each offset credit is equal to 1 metric ton of carbon dioxide equivalent ($MtCO_2e$) and can only be generated through implementation of an offset project for which the particular regulator has adopted a compliance-offset protocol. Once an offset is issued, it may be traded just like an allowance in the cap-and-trade program.

7.2.2   Regional markets

European Union Emission Trading Scheme (EU ETS): The European Union launched the EU ETS, the world's first government-mandated GHG cap-and-trade system, in 2005. The EU ETS covers 12,000 emitting facilities primarily in the power sector, specified industrial sectors, and combustion facilities with a thermal input greater than 20 MW. The ETS covers approximately 50 percent of EU CO_2 emissions covered

---

4 Under California's AB32 cap-and-trade program, regulated emitters can meet up to 8 percent of their triennial compliance obligation with offsets.

by the Kyoto Protocol. The goal of the EU ETS is to reduce emissions to 20 percent below 1990 levels by the end of 2020 (Monast *et al.*, 2009).

The EU ETS provides for the use of carbon offsets for partial compliance with carbon reduction obligations. Compliance entities may use certified emission reductions (CERs) created under Kyoto Protocol's Clean Development Mechanism (CDM) for up to 30 percent of their compliance obligation.

North American Carbon Markets: Throughout the U.S. and Canada, states and provinces are acting independently to create carbon reduction regimes aimed at reducing emissions, encouraging investment in clean energy technologies, creating green jobs, and improving public health.

- **California:** Under the direction of the Global Warming Solutions Act, known as AB 32, California has established the largest carbon cap-and-trade program in the United States. The program caps emissions from large industrial facilities emitting 25,000 $MtCO_2e$ or more, starting in 2013. In 2015, the program will also regulate emissions from the consumption of transportation and industrial fuels.

  The underlying regulation includes an enforceable GHG cap that will decline over time. During the initial phase of the program (2013–2014) the cap declines from 162 $MtCO_2e$ to 159 $MtCO_2e$. As transportation fuels are added to the program, in 2015, the cap increases to 394 $MtCO_2e$, but the program tightens the cap in successive years to 334 $MtCO_2e$ by 2020.

  The California program envisions a robust carbon offset program, but compliance entities are limited in the amount of carbon offsets they may use for compliance to 8 percent of their obligations. Currently, four categories of compliance offsets are accepted by the system's chief regulator, the California Air Resources Board (CARB). These include carbon projects created under the following offset methodologies established by the Climate Action Reserve:

  - Destruction of ozone depleting substances (ODS) originated in the United States
  - Destruction of methane from livestock manure
  - Sequestration of CO_2 from trees preserved through the execution of long-term conservation easements
  - Sequestration of CO_2 from the planting of trees in urban environments.

- **The Regional Greenhouse Gas Initiative (RGGI):** RGGI was implemented by ten states in the Northeastern U.S. in January 2009. RGGI covers fossil fuel power plants in the region with a generating capacity of at least 25 MW. The initial cap for the region is approximately 188 million tons of CO_2, declining by 10 percent between 2009 and 2018. The RGGI program is effectively administered on the state level, with each jurisdiction enforcing the cap on its regulated facilities. The RGGI program envisions a carbon offsets program, but credits can only be used at certain price levels. Due to an over-allocation of allowances relative to actual emissions, the cost of allowances in RGGI is low (>$2.00/short ton). The cost of abatement for carbon projects is considerably higher and therefore there has been limited use of offsets under the RGGI program.

- **Canada:** The province of Quebec has recently adopted legislation to implement a cap-and-trade system to fight climate change. Quebec's move was made only three days after the Canadian government's announcement that Canada will withdraw from the Kyoto Protocol. Under Quebec's new system, power plants are allocated emission rights under the cap and annual allocations diminish by one to two percent each year starting in 2015. The province of British Columbia is also considering the establishment of a carbon cap and trade program to accompany a carbon tax. This program is currently under review by the British Columbia provincial cabinet.

Chapter 8

Development: Development tools

Albie Fong

Because the solar industry has matured significantly since its infancy just a few years ago, site selection is becoming more of a science. Various developer tools are being utilized to save money for companies and encourage groups to make more informed decisions. If these tools were applied correctly, this would mean navigating through challenging areas without expending too many resources and recognizing fatal flaws earlier in the evaluation process. These developer tools are often data consisting of various maps delineating key project resources such as transmission lines or hazards, and areas to be avoided such as natural resource protection zones. Whether one is working in a mature renewable energy market such the southwest United States or Spain, or in emerging markets such as Mexico or Australia, obtaining access to this information is the first step. Additionally, regions or countries looking to foster the growth of such development often provide this information to assist in efficient development. Today, this information is increasingly available in digital formats and there are a number of software and mapping tools, both public and private, that developers should use to evaluate the viability of a given solar project site. In this chapter we will highlight examples of the types of developer tools available in the U.S. market and show how they have been applied to develop projects more efficiently. In other markets, developers should locate comparable information and apply them in a similar manner. If effectively used, project developers can more easily streamline early site review processes and efficiently filter through opportunities to determine the majority of development risks associated with a particular project. Finally, in this chapter we will present various databases and their respective viewing systems, but the reader is advised to use a Geographic Information System (GIS) software such as ArcGIS or even Google Earth to organize available data for coherent and repeated evaluation.

Typical tools that will be discussed are as follows and should be obtainable for any market a developer wishes to work in:

- Transmission Line Routes and Information
- Renewable Energy Resource Data (Hourly/Daily/Monthly/Annual Average Solar Radiation, Wind Speed, Geothermal Resource Temperature, Tidal Flow, etc.)
- Protected or Sensitive Natural Resource Area Maps
- Land Ownership Data

Figure 8.1 Screen Shot: U.S. natural wetlands development database.

8.1 USES AND APPLICATION OF GEOGRAPHIC INFORMATION SOFTWARE TOOLS: WETLAND DATABASES

Wetlands are protected areas of global interest as local water systems not only contain wildlife but are also portals to distant ecosystems and even drinking water. The U.S. Fish and Wildlife Service have a National Wetlands Inventory mapping database to indicate the most recent digital wetlands data available. In project development, there is never a definitive answer that ties a specific characteristic to the viability of the project. However, in the case of wetlands data, while there is not a clear definition of what is allowable for utility-scale projects, knowing what percentage of your site is classified as wetlands and the category of wetland (such as a 'freshwater emergent wetland' or 'estuarine' and 'marine deepwater wetland') will allow the developer to make a determination of the potential challenges associated with installing a solar field at a site. Perhaps with creative design and installation practices, minimal wetlands will be impacted.

A great visual example of the application of just this type of wetlands resource data is that of the 15 MWdc Jacksonville Solar project in Florida. Completed in 2010, it is a 91-acre site that needed to avoid wetlands to negate additional permitting time and project risk. The developer, Juwi solar Inc., most likely utilized available wetland maps and optimized the system layout, eventually utilizing a fixed array with solar panels to avoid wetland issues and their interaction with the system's operations. While some developers will use the National Wetlands Inventory data to avoid existing wetlands altogether, the Jacksonville Solar project used the information to create a strategic plan to navigate the wetland challenges as shown in Figure 8.2 and Figure 8.3.

Figure 8.2 Overhead view of Juwi solar Inc's Jacksonville Solar 15 MW solar project area and wetland areas (Collins, 2010).

Figure 8.3 View of Jacksonville Solar 15 MW solar project area and wetland areas. Note how the wetland areas were avoided in the final site layout (Collins, 2010).

8.2   BIOLOGICAL GEOGRAPHY DATABASES AND BASIS DATABASE NAVIGATION

If not utilizing a proprietary GIS System there are software tools available that have a much more diverse amount of data in one view to allow the user to evaluate multiple constraints at once. A good example of this is the Biogeographic Information & Observation System (BIOS), a system provided through the California Department of Fish & Game (DFG) along with some of its partners, which include the US Geological Survey, Bureau of Land Management (BLM), US Fish and Wildlife Service, California Coastal Conservancy, California Geological Survey, and US Forest Service Region 5. The purpose of BIOS is to explore the attributes and spatial distribution of biological organisms and biological systems studied by the DFG and its various partners who have provided data layers, or are working to provide layers, to BIOS. We are going to be discussing the features of the Public BIOS Data Viewer offered on the DFG BIOS website.

Upon initiating the BIOS, you will be presented with the view shown in Figure 8.4.

There are a number of base layers on the left window pane which the user can turn on and off depending on which features they are looking to evaluate with respect to a particular project site location.

Some of the major layers that are commonly used in the BIOS by project developers include:

- Hydrography
- National Wetlands Inventory
- California Protected Areas Database
- Public, Conservation, and Trust Lands

Figure 8.5 shows some unique features that can be seen in this area just outside of Bakersfield, California with the previously listed layers turned on. Users can simply select any of the selected layers on the left base layer pane and use the identify tool to

Figure 8.4 Screenshot California BIOS.

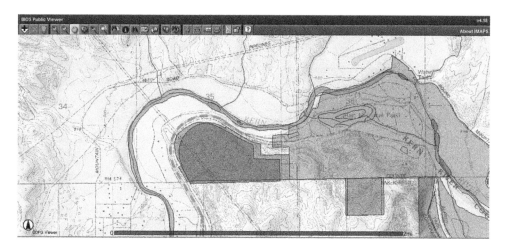

Figure 8.5 Land ownership delineation in California BIOS.

Figure 8.6 Screenshot BIOS County level view.

select a particular feature of that layer. The identify tool is used in Figure 8.5 to define the river feature as the Kern River, the dark shaded area in the middle of Figure 8.5 is 687 acres of land owned by the DFG, and the large land mass to the right of the DFG land is a 1,324 acre land that is administered by Kern River County Park. If one were to develop a solar project in this vicinity, understanding some of the nearby land ownership and lead agencies would assist in better understanding who to interface with, and understanding how your project would interfere with the operations of publicly owned land.

The different viewpoints allow for more or less detail depending what you are trying to capture.

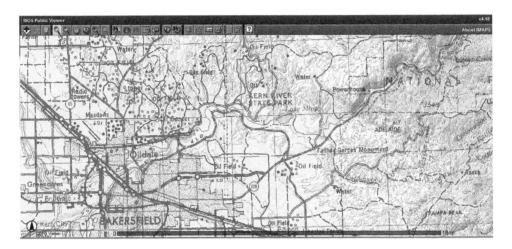

Figure 8.7 Screenshot BIOS city level view (with features such as gas & oil fields, radio towers, and water).

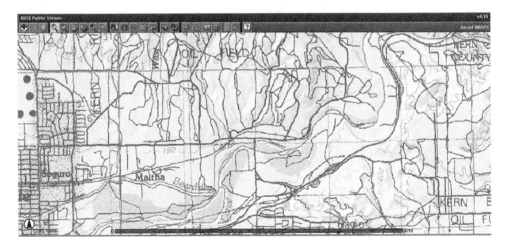

Figure 8.8 Parcel level view.

BIOS also provide additional layers that allow the user to customize his map. The black plus symbol in the BIOS toolbar area allows the user to add a layer. As of November 11, 2011, there were 359 available layers that are allowed to be added. This includes many layers that identify environmental zones of particular animal and plant species. By correlating important lists from other permitting agencies such as bird species of special concern from the department of fish and game, BLM areas of special concern, or a local entity such as wildlife corridors, restrictions on development can begin to be shown clearly.

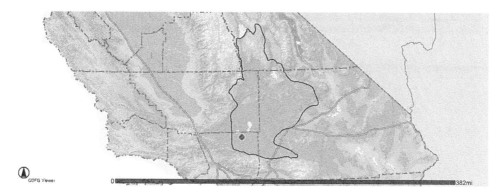

Figure 8.9 Screenshot BIOS showing sample project site and Mohave ground squirrel habitat (dark outline).

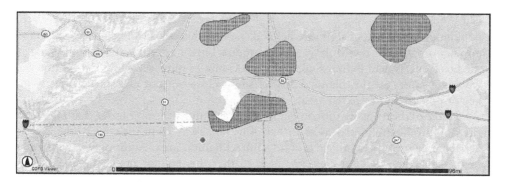

Figure 8.10 Screenshot BIOS habitat areas are defined by the circular outlines shown in the image above.

For example, let us imagine a project is proposed in Redman, California. It may be known, from reviewing other critical lists that the Mojave ground squirrel is of key concern in the Redman area. We may begin by reviewing this layer in relation to the planned project site. Utilizing this example in Figure 8.9, it can clearly be seen that the project site, identified by the diamond in the image center, is located within the southwest of the historic range, outlined in black, of the Mohave ground squirrel species. The historic range of the Mohave ground squirrel spans over several counties so viewing this data in Figure 8.10 gives us a better idea of where the project is situated within the range.

It is important also to view data layers at different scales; some information is only clear from a closer perspective. Some layers provide more detail as you zoom to a closer view of the site and the surrounding area. When getting a close view of the core area and population zone of the Mohave ground squirrel layer in the nearby vicinity of the Redman project site within Figure 8.11, you can easily see that the core area

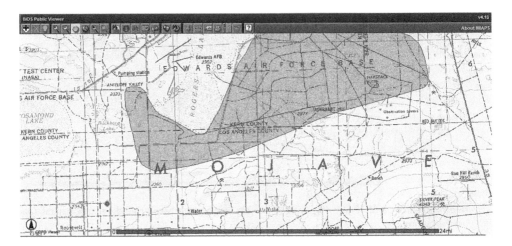

Figure 8.11 Screenshot BIOS showing project area, Edwards AFB and Mohave ground squirrel habitat.

population zones are much more defined than the historical range. There exists four population areas of the Mohave ground squirrel in an approximate 40-mile range of the Los Angeles, San Bernardino, and Kern Counties intersect. Hence, from this view we might note that this planned project site is close but not necessarily directly in a key or potentially restricted area.

Usage of the measure tool, in Figure 8.11, can allow the user to see that the project site is approximately 3.83 miles or 6,164 meters away from the edge of the 76,761-acre core area of the Mohave ground squirrel habitat area. Additionally, zooming in further on Figure 8.11 it can even be seen that this habitat range encompasses the Edwards Air Force Base, yet it is clear that at least for this species, the project area is outside of the project habitat.

Doing this sort of "desktop" review of a project area before committing large amounts of funds can not only save money but also time. No matter what area of the world one is working in these types of maps exist and should be sought out. With today's electronic information you can discover in 30 minutes what would have taken hours of "on the ground" research to uncover.

8.3   FLORA RESOURCE INFORMATION DATABASES

Addressing wild California plants that can be key species under conservation is Calflora. It has a database of information for over 800,000 plant locations (based on observations) including more than 10,000 native and introduced species. The Calflora Database is maintained by a nonprofit organization, which provides information for the purpose of education, research, and general conservation efforts. Since Calflora relies on information from contributors, it will include data ranging from high-level plant research professionals, California environmental consultants, to everyday amateur plant enthusiasts. Solar project developers can utilize Calflora to take advantage of the wealth of plant observation location data to determine what kind of sensitive environmental habitat may have a presence in the nearby vicinity of a targeted project site. As the data is basically open source, developers have to use the information wisely and correlate with other environmental information that applies to a specific region or site. Calflora should be used as a tool to make more educated decisions about the existence of species that may hinder permitting and the eventual construction of the project.

If there is one particular plant species that the developer is seeking more detailed information about, there is a search feature on the website where you can filter the plant name by common name, scientific name, or family name. Other search filters include life form characteristics, native and non-native distinction, livable elevation criteria, community type, and by county.

Calflora example plant search

After doing a search for the California jewelflower, a general description page of the species appears with example photos. Table 8.1 indicates some of the general information provided for this endangered species.

Below the general description information, a distribution map (Figure 8.12) identifies counties in California where the California jewelflower exists with one or more occurrence records within a single county. For this particular species, there are seven different California counties where there exists some kind of documented observation.

Below the California description map, the total amount of records of the California jewelflower is clearly identified (Table 8.2). There are 210 records among seven bordering counties in the California central valley. Within the total records, some of the records are Consortium of California Herbaraia (CCH) records, which identify specimens housed in herbaria. The number of CCH records is separately differentiated from the total records for the user's reference.

To gain a more detailed distribution grid of the California jewelflower, a Google map can be viewed which will show the rectangular quadrants where the records exist within the county. Figure 8.13 obviously gives a more detailed observation point for the user to see the California jewelflower distribution better. The closer view in Google Maps, the more detailed the quadrants become so it's up to the user to discern which view is best for what they are seeking to evaluate.

The Calflora Map Viewer indicates three different types of observations: specimen, reported, and literature. A specimen observation is defined as a specimen record that serves as a basis of study and is retained as a reference in an accessible

Table 8.1 CalFlora example search.

Species	California jewelflower (caulanthus californicus)
General Characteristics	General characteristics: annual herb native to California and endemic (limited) species to California
Protected Status	Endangered in the state of California and by the Federal Government
Elevation	Between 0 to 3,000 feet

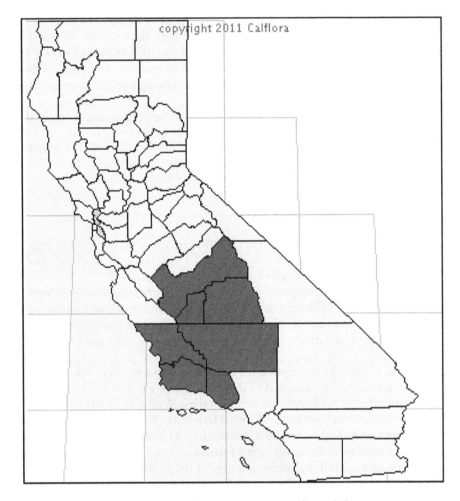

Figure 8.12 Calflora showing california jewelflower habitat.

Table 8.2 California Jewelflower records by region.

County	Total records
Fresno	21
Tulare	20
Kings	11
Ventura	6
Kern	76
Santa Barbara	23
San Luis Obispo	53

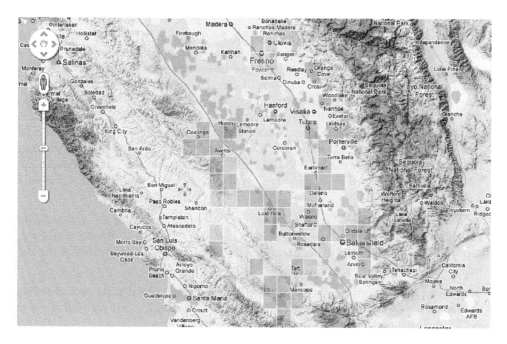

Figure 8.13 Screenshot CalFlora Map Viewer showing habitat areas.

collection. A reported observation is a record where the observer, date, and location are all known. A literature observation is an indirect report. Literature observations could include a record from a literature source that describes an area in general terms or where the observer is not known but the information was aggregated from a reliable source. Record details can easily be viewed by clicking on the observation icons.

While it is interesting to know that a California jewelflower specimen observation was retained in 1935, this obviously doesn't serve as a basis for the current health and conservation status of the species. However, resources like Calflora could be used as an effective tool to determine if further evaluation needs to be conducted in a fatal flaw analysis for a species of a particular site. If there are an abundance of observations that congregate around a particular site or region of interest, that may point to something that the developer should look into with other environmental consultation or interface with the permitting authority.

8.4   RENEWABLE ENERGY RESOURCE INFORMATION: SOLAR RADIATION DATA

With other useful layers in BIOS, users can evaluate a lot of different features relating to solar development. This is an excellent example of combining data sets to answer related questions such as; "Where is a good place for a solar project avoiding habitats?" The BIOS discussed here uses the solar resource data made available by the National Renewable Energy Lab (NREL) to evaluate both direct normal radiation and global horizontal radiation. Figure 8.14 gives an impression of what the direct normal radiation looks like around the Redman site and, there are four distinct regions of direct normal radiation with the darker shades representing the areas with more abundant solar resources (see Figure 8.15 for a legend of the different values stated in watt hours per meter squared per day). When using the identify tool to select a particular location, the following data will appear under the BIOS map, which includes: latitude, longitude, average monthly values (DNI03 equals 7,128 W/m²/day in the March), and yearly average value (this location has a yearly average direct normal radiation of 7,488 W/m²/day).

Tip: Sometimes turning on multiple layers will not allow the user to evaluate different characteristics at the same time. It can be advantageous to turn on layers separately and mix and match layers to see which ones can be effectively used at the same time.

Tip: If you need to print an image such as a jpeg direct normal solar radiation map, instead of solely accessing them via a software based tool, you can find it on the NREL website: http://www.nrel.gov/csp/maps.html#nm

   While having access to the solar radiation data through the BIOS is useful when comparing it against other environmental characteristics, you have even more flexibility to evaluate the solar radiation characteristics with stand-alone data such as those available through Solar Prospector. The Prospector is a solar resource mapping tool developed by NREL to aid developers in siting utility-scale solar projects. The detailed map layers, downloading of data, and solar resource analysis are the primary features that allow the user to filter through different project sites. The Prospector is a great public desktop analysis tool where large amounts of evaluation can occur without visiting the project site.

   Compared to Figure 8.14, which shows a similar solar resource feature but on the DFG's BIOS, Figure 8.16 is more user friendly as it is overlaid onto a Google Maps interface.

Tip: transparency of the various layers can be easily modified within the options of the particular layer.

Tip: unlike DFG's BIOS, the user can use the Prospector to filter a particular DNI (Direct Normal Incidence – a measure of solar energy at a location on the earth's surface) level to look for areas with advantageous solar resources. Multiple ranges could be selected at the same time to widen the search criteria. In this scenario, areas with annual average DNI of 6–6.5 kWh/m²/day were filtered (Figure 8.18).

Tip: Prospector can easily allow the user to locate sites with multiple features. In Figure 8.19, two layers were selected: one identifying solar DNI areas with 6–6.5 kWh/m²/day (light shaded areas) and the other identifying areas with less than 1% land slope (checkered dark areas).

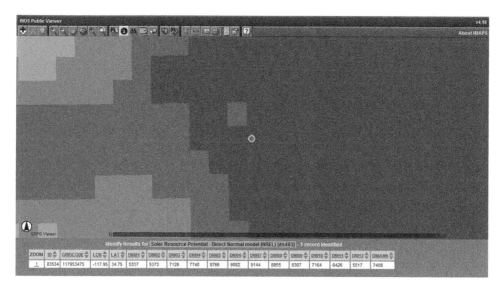

Figure 8.14 Screen Shot BIOS showing solar radiation.

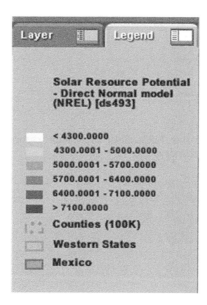

Figure 8.15 Solar radiation legend.

While Prospector is primarily used for its ease in identifying solar resource characteristics, it also provides additional layers like BIOS, which can be used independently to evaluate other characteristics. Generally Prospector is more user friendly than BIOS and loads the layers quicker. One of the useful layers includes the environmental layer,

Figure 8.16 Screenshot NREL Solar Prospector showing solar radiation intensity near Bakersfield, CA.

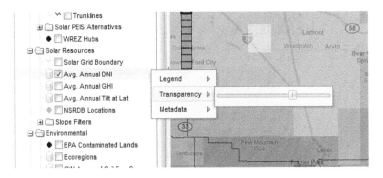

Figure 8.17 Screenshot NREL Solar Prospector, changing transparency of radiation data.

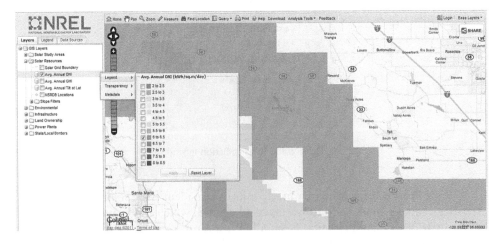

Figure 8.18 Screenshot NREL Solar Power Prospector showing radiation legend.

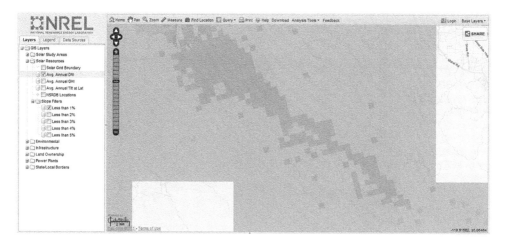

Figure 8.19 Screenshot NREL Solar Power Prospector showing filtered data.

which has data indicating the Southwest areas of critical environmental concern, the Southwest areas of fauna critical habitat, and the Southwest areas of flora critical habitat. BIOS provides layers that are more specific than those in the Prospector since it provides layers specifically for particular types of endangered animal or plant species. Each layer comes from a different source, which can be checked under the Metadata options. For instance, the user will easily be able to find out that the Southwest area of critical environmental concern data is from the BLM & Argonne National Laboratory and the Southwest fauna critical habitat data is from the U.S. Fish and Wildlife Service & Argonne National Laboratory. Other layers easily identify lands that are federally owned, tribal lands, related information to the Renewable Energy Transmission Initiative (RETI), and the slope of the land. These are just some examples, the key for the developer is finding the best sources that are easy and efficient to use in the desired region.

Another brief example within the Solar Prospector tool is the following; Figure 8.20 shows the areas in the desert Southwest that have critical habitats for fauna and critical environmental concerns. Potential solar project sites that are situated in both of these areas should be avoided to reduce potential interference with conservation efforts. It is not impossible to develop in these areas, but more extensive biological evaluations may need to be conducted during the permitting phase, and some potentially expensive mitigation efforts may need to occur to allow the solar project and existing species to co-exist.

Specific market opportunities such as the use of solar thermal power with conventional power plants can also be investigated. Markers for existing coal plants and natural gas combined cycle plants can be placed onto the Prospector map to identify potential candidates for solar steam augmentation projects, which could benefit groups developing solar thermal projects. Figure 8.21 represents a layer for areas that have more than a 6 kWh/m^2/day DNI, overlaid with triangular markers representing the location of the natural gas combined cycle plants. Natural gas combined cycle

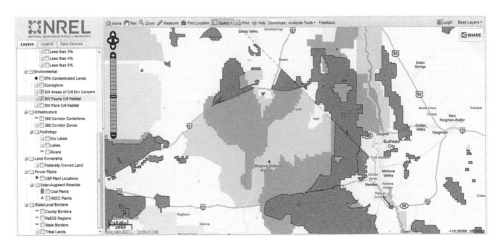

Figure 8.20 Screenshot NREL Solar Power Prospector showing example environmental concern layers.

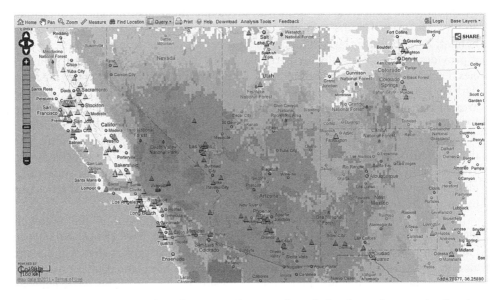

Figure 8.21 Screenshot NREL Solar Prospector showing solar radiation data and power plant locations.

facilities that fall outside of the high DNI areas would have a lower likelihood of solar augmentation making economic sense due to less annual generation that would occur for any given solar field of solar thermal equipment.

Using the query tool (by point) from the toolbar, the user can easily find the average monthly and yearly DNI for a particular rectangular quadrant as depicted in Figure 8.22.

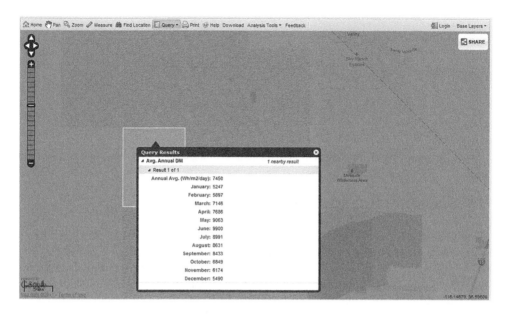

Figure 8.22 Screenshot NREL Solar Power Prospector showing query results for specific data.

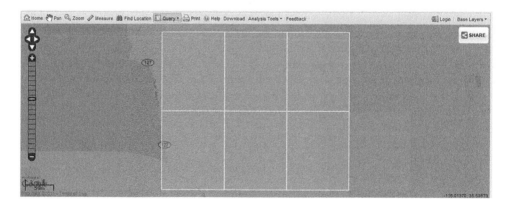

Figure 8.23 Screenshot NREL Solar Power Prospector showing selection of area with 'query tool'.

Using the query tool (by region), the user can select multiple quadrants to contrast and compare monthly and annual DNI averages as depicted in Figure 8.23 and Figure 8.24. In this case, six quadrants have been selected and there are obvious deviations in the DNI values with almost 500 Wh/m²/day difference between the quadrant with the lowest average annual DNI compared to the highest.

Avg. Annual DNI [×]												
Download Results ✕ Close All Results												
Annual Avg. (Wh/m2/day)	January	February	March	April	May	June	July	August	September	October	November	December
7484	5436	5787	7227	7515	8946	9954	9117	8730	8523	6849	6201	5436
7429	5373	5724	7002	7461	8937	9945	9072	8631	8442	6786	6255	5436
7292.1744	5626.908	5802.678	7062.8094	7089.1632	8605.9584	9575.4411	8779.428	8337.7755	8085.7944	6900.8589	6124.275	5433.5907
7469	5391	5778	7128	7479	8973	9927	9144	8622	8451	6921	6273	5454
7750	5535	5886	7344	7839	9297	10332	9495	8973	8703	7128	6588	5778
7475	5346	5679	7218	7560	8955	9981	9180	8649	8451	6939	6228	5418

Figure 8.24 Screen Shot NREL Solar Power Prospector showing table output from regional 'query tool'.

8.5 GEOGRAPHIC SOIL DATA

One characteristic that BIOS and Prospector do not cover is soil data and associated information that can be important for evaluating construction considerations. By utilizing the Web Soil Survey (WSS) software from the United States Department of Agriculture (USDA) Natural Resource Conservation Service (NRCS), developers can gain valuable information about the soil characteristics that would be similar to a geotechnical report. This can help to determine what kind of design challenges might be present for the purpose of different structures that would eventually be placed on the project site. For instance, single axis and dual axis trackers have different structural and soil requirements hence the soil type may become an important variable in system cost estimation. Soil tests should be conducted at later stages of development however, software tools like the WSS provides a sufficient analysis prior to spending a significant amount of development capital to make good preliminary decisions.

The following is detailed information regarding WSS. The reader is reminded that this web-based software is only valid in the U.S. however, in any part of the world, similar resources may exist and should be sought out. Initial review of data such as this in the early stages of a project will save time and money.

There are four main tabs associated with the WSS:

1) Area of Interest (AOI)
2) Soil Map
3) Soil Data Explorer
4) Shopping Cart

Area of Interest (AOI) – To start the soil evaluation process through WSS, the user simply defines the location to be evaluated by defining the latitude & longitude, Public Land Survey System (PLSS) section/township/range, specific address (if applicable), or one of the other features provided. Generally, defining the latitude & longitude or section/township/range is the easiest way to navigate through the WSS, as it will bring up the general area around the site. Then the area to be evaluated needs to be defined, which can be done with the "Define AOI by rectangle" or "Define AOI by polygon" tool.

Tip: If you define an AOI by the latitude and longitude, you can find the nearby section/township/range by selecting the "Legend" tab to the left of the toolbar and

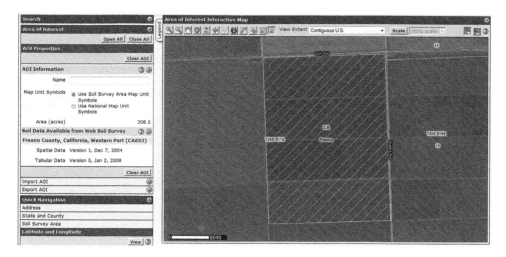

Figure 8.25 Screen Shot WSS Soil Data Explorer showing selection of an Area of Interest.

then choosing the "PLSS Township and Range" option. Choosing the "PLSS Section" option shows the square mile section of land, which is 1/36 of a township.

In Figure 8.26 we defined the area of interest, which was 208.5 acres as defined in the "AOI Properties" on the left toolbar.

The "Soil Map" tab will identify the different types of the soil within the AOI with an associated unit number to identify the soil type. In this case, 92.6 acres are Westhaven Loam, 0–2 percent slope identified with the number 474, and 115.9 acres are Westhaven Clay Loam, 0–2 percent slope identified with the number 477.

The Soil Data Explorer tab gets into detail of the suitability and limitation ratings of the area. There are eleven different sub-topics, which contain numerous sub-topics underneath them. The user needs to determine which areas are most relevant to their soil research goals. The various filters provide information on the suitability of the soil for a given use such as roads, fences, or buildings.

In Figure 8.27 the Small Commercial Buildings filter has been used as it highlights the critical soil issues related to constructability. In this scenario, it can be seen that the entire site has a type of limitations or issues that defined in Figure 8.28. In this case the soils at this site expand and contract with moisture potentially requiring additional engineering to the foundation of a given structure. In the ideal scenario, the entire site, if not the majority, will have a green coloring for the AOI, which indicates no limitations. This tool allows for a rapid analysis of a wide land area without in depth civil or geotechnical expertise.

The Shopping Cart tab allows the user to customize a printable report, in PDF format, with the results found in the Soil Data Explorer and Soil Map. Included in this report are the maps generated in the previous tabs along with further details like elevation, mean annual precipitation, mean annual air temperature, frost free period (in days), slope, depth to water table (in inches), depth to restrictive feature

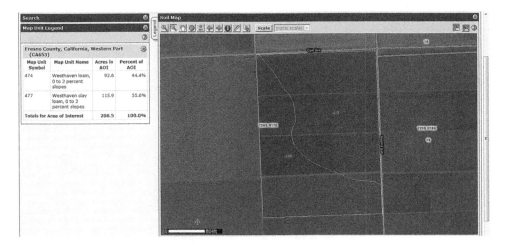

Figure 8.26 Screen Shot WSS Soil Data Explorer showing soil data.

Figure 8.27 Screen Shot WSS Soil Data Explorer showing constructability ratings.

(in inches), frequency of flooding, minor components with their associated percentages, and many other interesting soil related features. Automatically generated reports like these are rapidly generated and provide excellent information prior to site visits. They could also be used for discussions with civil or geotechnical experts.

Map unit symbol	Map unit name	Rating	Component name (percent)	Rating reasons (numeric values)	Acres in AOI	Percent of AOI
474	Westhaven loam, 0 to 2 percent slopes	Limitations	Westhaven, loam (85%)	Shrink-swell (LEP 3-6) (0.22)	92.6	44.4%
477	Westhaven clay loam, 0 to 2 percent slopes	Limitations	Westhaven, clay loam (85%)	Shrink-swell (LEP 3-6) (0.22)	115.9	55.6%
Totals for Area of Interest					208.5	100.0%

Tables — Small Commercial Buildings (CA) — Summary By Map Unit

Summary by Map Unit — Fresno County, California, Western Part (CA653)

Figure 8.28 Screen Shot WSS Soil Data Explorer showing data related to soil constructability in the AOI.

8.6 TRANSMISSION ROUTE AND INFORMATION DATABASES

While many software developer tools are widely available for public usage, tools that often have commercial uses come with a cost. One extremely useful, but not public, database available from Platts is the Platts Electric Substation geospatial data layer or the Platts Transmission geospatial data layer. For the purpose of evaluating the local transmission and substation infrastructure with respect to a particular region (presently Platts covers North America and parts of Europe) or project site, this data will be sufficient for pre-development needs relating to project siting and interconnection work (see Figure 8.29).

Figure 8.30 depicts a view of both the substation and transmission infrastructure layered together in the area of Havasu, Arizona.

Platt's GIS data has been found, for the most part, to be extremely accurate with few occurrences of inaccuracies. Furthermore, a developer can often distinguish potential inaccuracies by correlating where Platts indicates an existing transmission line and utilizing close up views through satellite view in Google Earth or Bing Maps to see the actual transmission line towers. Figure 8.31 is from Google Earth and shows an 115 kV transmission line to the east of the First Solar 21 MWe solar farm in the city of Blythe. There is also a 500 kV line that runs through the project site.

To verify that there is actually 115 kV transmission infrastructure east of the solar facility, Google Maps in satellite view depicts five transmission towers south of 7th Ave. This provides a sanity check that the Platts GIS data is accurate and that if you conduct a site visit, you will see and be able to verify those key transmission features.

Additionally, Google Street Views when available can not only confirm locations of transmission lines, but also show the type of vegetation and slope of the land as seen in Figure 8.32.

Since the substation and transmission line data come as separate metadata, you can easily decide to turn one or the other off depending on the particular feature you are looking for. Some developers look for locations that are geographically within a range of various different points of interconnection and perhaps different substations. Increasing the amount of transmission lines or substations increases the possibility of existing free capacity to offload the solar energy generated. At the same time, being in that kind of area also increases the likelihood of a significant amount of other projects

Figure 8.29 Sample Arizona and California transmission lines (Transmission data courtesy of Platts).

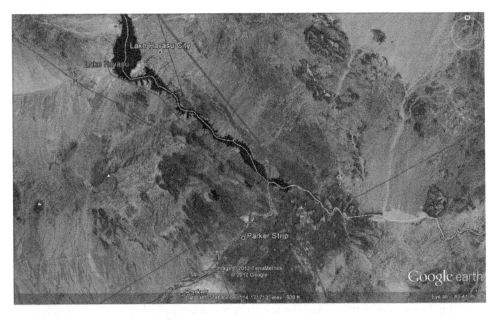

Figure 8.30 Lake Havasu area transmission infrastructure (Transmission data courtesy of Platts).

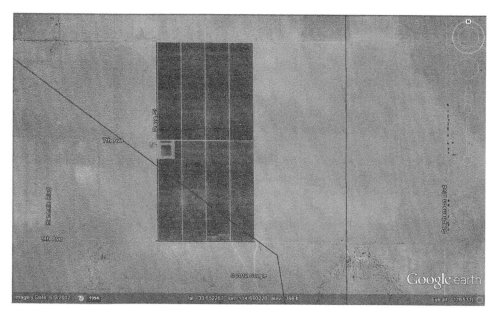

Figure 8.31 First Solar's 21 MWe Blythe Solar farm shown with area transmission infrastructure (Transmission data courtesy of Platts).

Figure 8.32 Screenshot Google Maps Street View showing transmission or distribution lines.

in the nearby vicinity; leading to potentially expensive network upgrade costs in order for all generators to safely dispatch their energy. Being too far from any existing transmission infrastructure would likely lead to the cost of building new transmission lines outweighing the benefit of the revenue generated from the solar project. While there is no outright specific type of place to site a project, all variables have to be taken into account with respect to the costs of transmission, and GIS data provided from Platts can help the user make a more calculated decision.

Suggested important questions to answer when conducting transmission infrastructure reviews for a project are:

a) How far is the nearest existing transmission line and what is the voltage level?
b) How far is the nearest substation and what is its rating?
c) Do any of the interconnection infrastructure run through the site or adjacent to the site for easy access?
d) If there is interconnection infrastructure on the target site, is there enough suitable building space to develop the intended project size?
e) How many other projects are intending to interconnect to the same point of interconnection or nearby substation?
f) To access the nearby substation or transmission line, are any right of way permits required?

Chapter 9

Development: Design considerations of photovoltaic systems

Alfonso Tovar

As with most engineering projects, the design, construction and implementation of a photovoltaic power plant is a multi-disciplinary effort that involves very diverse knowledge areas. Throughout this book, several of these components have been reviewed, including legal and financial aspects. This chapter will cover the general concepts required to design large-scale photovoltaic power plants. The content of this chapter is focused on the practicalities of the design of a photovoltaic (PV) power plant and is intended for a non-specialized audience. The discussion of all the topics in this chapter is qualitative in nature, covering the engineering principles without developing a detailed scientific analysis or engineering assessment. Several books have already been written with that emphasis and the intention of this book is to present a general perspective with enough depth to understand the fundamentals of the system.

9.1 OVERVIEW OF DESIGN CONSIDERATIONS

A photovoltaic system is an electrical energy generator fueled by sunlight. In contrast to other power generation technologies, photovoltaic systems are based on discrete power generation units: photovoltaic modules. The power range of most photovoltaic modules available in the current market is from 30 Watts to 300 Watts, and, because each module is a complete power generator, a photovoltaic system can consist of one single module or the aggregated capacity of many modules[5]. The focus in this book is on applications larger than 10 Megawatts (MW) of power, also referred to as utility-scale photovoltaic systems or photovoltaic power plants. Although these terms have not been standardized, the common understanding is that besides the large size of these systems, they also connect directly to transmission systems.

In addition to the module, a PV plant is integrated by other components, the most relevant of which are inverters, mounting structures, monitoring and control systems, wiring, electrical protection gear (fuses, breakers, lighting arrestors, etc), energy storage units, and transformers. The main characteristics and functions of these components will be discussed throughout this chapter.

---

5 The largest photovoltaic system in the world at the end of 2011 was rated at 80 Megawatts (MW). Over one million modules were used to complete this power plant. Several projects under development will at least double this capacity before 2015.

The integration of these components into an efficient system is the final result of a design process that fundamentally requires information about the following to start:

1) Solar resources and weather patterns on site
2) Photovoltaic module selection
3) Specific project requirements
4) Photovoltaic inverter selection
5) Module mounting structure selection

1) The solar resources available at the project's site will determine the amount and distribution of solar irradiance expected at the location of interest. Commercial photovoltaic technologies have different performance responses depending on the qualities of the solar resource, and this will define the maximum potential energy that the PV plant can generate. In addition to the solar irradiance pattern throughout the year, other environmental patterns, most significantly the ambient temperature, wind speed, precipitation, snowfall, dust and pollution, will also affect photovoltaic systems' performances. It is important to note that the selection of photovoltaic technology is generally based on the project's economics and not solely on the basis of best performance given the solar resources and environmental patterns on site.

2) The modules are the single most expensive item of total project costs and thus, the selection of the specific photovoltaic technology (crystalline silicon, thin-film, or high concentrating systems) and vendor is often tied to the financial analysis of the project. However, there are some performance advantages and disadvantages to consider, depending on the technology, as well as various stages of commercial maturity and degrees of confidence on the part of the investor about specific technologies or manufacturers. The selection of the photovoltaic module is not strictly based on component price.

3) Each project has to address specific requirements determined by the site conditions (e.g. topography, soils, drainage) and infrastructure available (e.g. access roads, points of interconnection, transmission lines). Often, there are also a number of specific performance requirements for the PV plant such as seasonal or daily production profiles, operational and interconnection features, maximum alternating current (AC) or direct current (DC) capacity, and others. These performance requirements may be defined by the developer, the owner, the utility and/or the grid operator that the PV plant is connecting to.

4) The selection of the inverter is typically independent of the module selection with few exceptional cases. The selection of the inverter is mostly dependent on the performance requirements of the project. In addition to these features, the designer also needs to consider the overall characteristics of the inverter, expected long term performance, the manufacturer's experience with inverters, expected reliability, installation procedures, maintenance expenses, and, of course, equipment price.

5) The mounting structure is a component that the designer should also consider early in the design process, although only a couple of generic characteristics are required to complete the conceptual design. However, the complete features of the mounting structure are critical to advancing from the conceptual design to

the next phase. This is because the installation of the mounting structure depends on the specific soil conditions found on site as well as structural requirements determined by environmental loads (e.g. wind, snow, seismic activity). As a consequence, geotechnical and structural analyses will narrow the foundation and rack tilt options within a certain cost range. The characteristics of the mounting structure will define the energy generation profile, the project layout, and site preparation requirements. The mounting structure is sometimes considered during the early design phases, which can create inefficiencies in the design process by requiring design adjustments at more advanced stages.

The designer needs to analyze how the specific characteristics of each of these components interacts, and use this information to optimize those interactions with the goal of developing the most efficient design that best matches the project requirements. The first two steps recommended in the design process are to systematically organize all of the data available, and develop a design protocol or work plan that clearly identifies goals, tasks, sequences, people and responsibilities, timelines, information available, and information gaps. A typical design process involves a number of iterations between these different variables as illustrated in Figure 9.1.

9.2 SOLAR RESOURCE

The amount and distribution of the solar resource available on site is fundamental to estimate the electrical energy that the system is expected to generate. For most large scale commercial systems in North America the analysis of the energy expected to be produced by the system is critical because it often defines the terms of project finance[6]. The electrical energy is estimated through a number of calculations that are best performed using a software program. There are a number of commercially available computer programs that can be used to build a mathematical model of the photovoltaic system in order to estimate the total energy output generated by this system. Most software programs are based on the single diode model of a photovoltaic cell and single point efficiency of an inverter, although there are some programs based on parametric equations. There is a vast literature available on photovoltaic system models and performance for the interested reader. Regardless of the analytical approach used, the fundamental inputs required to calculate the energy production of a photovoltaic system are the solar resource, the characteristics of the photovoltaic module, the characteristics of the inverter, and the orientation of the system[7]. In addition to these inputs, other parameters need to be considered in order to generate a more accurate model and response of the photovoltaic system. The four inputs mentioned are the first order variables and the effects of them will be discussed throughout this chapter.

---

6 This is due to the specific context of solar market in this region. In other contexts (e.g. certain Feed-In-Tariff environments) the expected energy output of the system may not require a detailed analysis.

7 These minimal inputs are required for grid-tied systems. Off-grid or DC-DC systems may require additional inputs such as storage type, converter characteristics, etc.

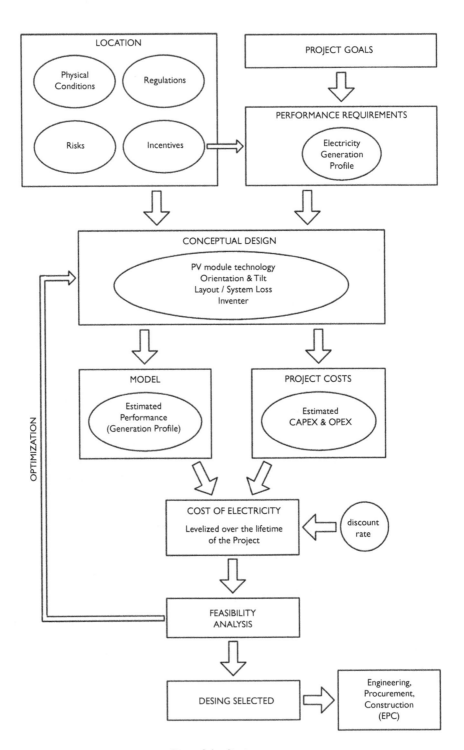

Figure 9.1   Design process.

The solar resource is the "fuel" of a photovoltaic system and thus, the amount of sunlight incident on the photovoltaic modules is the main input to calculate the electrical energy output generated by the system. There are three main characteristics of the solar resource that are important for photovoltaic applications:

1) Direction or angle of incidence
2) Atmospheric disturbance or clearness index
3) Inter-Annual Variability

9.2.1 Direction or angle of incidence

The direction of the sunlight depends on the position of the sun relative to the photovoltaic modules and on the degree of disturbance that the solar radiation suffers while traveling through the atmosphere from space to the Earth's surface.

The apparent movement of the sun through the sky as seen from a fixed location on the Earth's surface can be determined analytically with high precision[8]. Using a horizontal plane as a reference (refer to Figure 9.2), the sun travels East to West from morning to evening. The angle between the sun's position and the geographical East or geographical West is the azimuth. Typically, the angle is negative towards the East, zero at the reference, and positive towards the West. Using this convention, an angle of −90 degrees from the location is East and +90 degrees is West while 0 degrees would be true South or North, depending if the location is in the northern or southern hemisphere[9]. The sun rises over the horizon through the day and the angle between the horizontal plane and the sun is the elevation. The sun reaches its maximum elevation in the day at solar noon, which more or less corresponds to 12:00 pm, depending on the distance of the location to the time zone meridian and daylight saving adjustments. The azimuth and the elevation angles define the exact position of the sun as seen at the location. The zenith is an imaginary vertical line at the location or, in other words, a vector with a 90 degree elevation angle. The zenith is the complementary angle of the elevation.

In solar engineering it is important to note the effect of the location's latitude on the sun's position. The Earth has three main latitude circles:

- The Equator, at 0 degrees latitude, divides the Earth in the Northern and Southern hemispheres
- The Tropic of Cancer, located approximately 23.5 degrees north of the Equator
- The Tropic of Capricorn, located approximately 23.5 degrees south of the Equator

The maximum elevation of the sun as well as the range of the azimuth angle change throughout the year. In locations north of the Tropic of Cancer, the maximum elevation the sun can reach in the year is always less than 90 degrees. The trajectory

8 For example, "*Solar Position Algorithm for Solar Radiation Applications*" Reda, I., Andreas A. (2003). NREL Report TP-560-34302 Revised January 2008.
9 This widely used reference system presents some challenges for locations closer to the Equator, as it will be explained later.

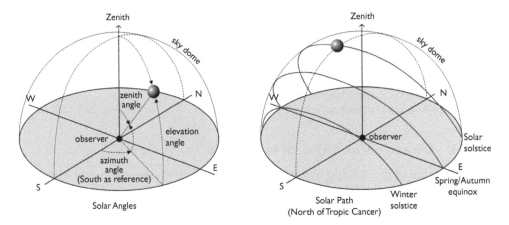

Figure 9.2 Solar angles and position.

of the sun will be an East to West arc tilted to the geographical South. In summer solstice, the sun reaches an elevation of 90 degrees at the Tropic of Cancer, or a zenith of 0 degrees at solar noon. If the latitude of the location is known, the maximum zenith angle of the sun can be easily estimated by subtracting 23.5 degrees from the latitude. The maximum elevation is the complement of the zenith angle, or 90 degrees minus the zenith angle. For example, the city of San Francisco in the United States is located at 37.75 degrees latitude. Therefore, the maximum elevation of the sun at this location will be ~75.8 degrees at solar noon on summer solstice (June 20 or 21 depending on the year)[10]. Before and after this date, the elevation of the sun will be always lower. The range of the azimuth angle will also be at its maximum in summer solstice. A rough estimate of the azimuth range consists in adding 85 if the location is between 30 and 36 degrees latitude, and 80 if the location is between 38 and 60 degrees latitude[11]. Therefore the approximate azimuth angle range in summer solstice for San Francisco will be −118 degrees from true South in the morning to +118 degrees from true South in the evening, making this day the longest of the year[12].

The lowest elevation of the sun for locations in the northern hemisphere occurs in the winter solstice (December 21 or 22 depending on the year). In this case, the minimum elevation can be estimated by subtracting the latitude from 66.5[13]. Therefore, the elevation of the sun in San Francisco at solar noon on winter solstice will be ~28.8 degrees. This is also the same as subtracting 47 degrees from the maximum elevation. 47 degrees is the distance between the Tropics and the maximum change in solar elevation within a year for locations north of the Tropic of Cancer.

10 Maximum elevation = 90 − (37.75–23.5) = 75.75.
11 Higher latitudes have a much wider azimuth angle and beyond 68 degrees the sun doesn't set on this day.
12 The actual azimuth angles are +/−120.2 degrees from true South.
13 Which is derived from: 90 − zenith = 90 − (location's degrees higher than Tropic of Cancer (23.5) + sun's position at Tropic of Capricorn (−23.5) = 90 − [(Latitude − 23.5) + (23.5 + 23.5)]. Locations north of 66.5 degrees latitude are within the Arctic circle and will not see the sun this day.

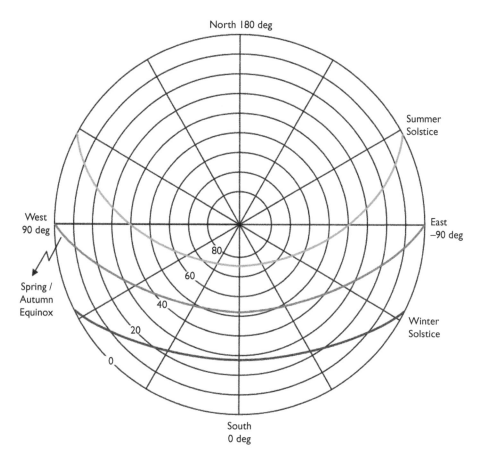

Figure 9.3 Solar position – San Francisco.

The range of the azimuth angle will be at its minimum in the winter solstice and the approximate range can be found by subtracting the latitude from 95 if the location is between 30 and 36 degrees latitude, and 100 if the location is between 38 and 60 degrees latitude. Thus, for San Francisco, the approximate azimuth angle range on the day of winter solstice is +/−62 degrees from true South, which makes this day the shortest of the year. Figure 9.3 is a two dimensional plane that illustrates the sun's trajectory in the summer and winter solstice as seen in San Francisco.

The summer and winter solstices define the extreme trajectories of the sun, which for San Francisco, means a maximum elevation of 75.8 degrees, a minimum elevation of 28.8 degrees, a maximum azimuth of +/−118 degrees and a minimum azimuth of +/−62 degrees. The elevation and azimuth of the sun change daily between these two extremes in the year. The spring and autumn equinoxes provide two more references in time to quickly estimate the sun's trajectory. On these two days the sun reaches an elevation of 90 degrees at the Equator, or a zenith of 0 degrees at solar noon. From the location's perspective, if the sun is over the Equator, the zenith angle of the sun, for example in San Francisco, is simply the angle difference between the Equator and San Francisco's latitude, or 37.75 degrees. Thus, the elevation of the sun is the complementary angle to the zenith angle, or 52.25 degrees. From summer to winter, the elevation of the sun is decreasing every day and is half its maximum on autumn equinox (September 22 or 23, depending on the year). From winter to summer, the elevation of the sun is increasing every day and is twice its minimum on spring equinox (March 20 or 21 depending on the year). The azimuth angles are +/–90 in both cases.

In locations south of the Tropic of Capricorn, the elevation and azimuth angles can be estimated in the same manner as explained above, but in the southern hemisphere, the maximum elevation of the sun will occur on winter's solstice day and this will also be the longest day of the year. The lowest elevation of the sun at solar noon will occur the day of summer solstice and this will be the shortest day of the year. In the southern hemisphere the trajectory of the sun will be an East to West arc tilted to the geographical North. By convention, southern latitudes use a negative number and the calculations provided will have to be adjusted for this. The spring and autumn equinoxes will be the same in both hemispheres, with the exception of the tilt of the East to West arc.

For locations in between the Tropics, the sun's trajectory is different in terms of the maximum elevation, the tilt of the East to West arc and the extent of the azimuth angles. A location along the Equator will experience the most significant changes. This case will be used to illustrate the general characteristics of the sun's trajectory within these latitudes. For example, the city of Quito in Ecuador is located at –0.2 degrees latitude, barely in the southern hemisphere. In this case, the elevation of the sun will reach a 90 degree angle two times in the year, during the spring and autumn equinoxes. The azimuth angle in these two days will be +/–90 degrees. The tilt of the East to West arc is to the North from spring equinox to autumn equinox and the minimum elevation of the sun is on Summer solstice when the sun is over the Tropic of Cancer. The minimum zenith angle can be calculated by subtracting the location's latitude from 23.5 degrees, noting negative values for southern hemisphere latitudes. Thus, for Quito, the minimum zenith angle is 23.7 degrees and the minimum elevation of the sun at solar noon is 66.3 degrees. During the other half of the year, from autumn equinox to Spring Equinox, the tilt of the East to West arc is to the South and the minimum elevation of the sun is on Winter solstice. The minimum zenith angle is also calculated by subtracting the location's latitude from –23.5 degrees. For Quito, the minimum zenith angle on Winter solstice is 23.3 degrees or the minimum elevation of the sun at solar noon is 66.7 degrees. For practical purposes the azimuth angle range can be considered independent of the location's latitude and the same applies for any place between 0 degrees and +/–25 degrees latitude. The use of a chart with negative and positive angles can be confusing for locations within the Tropics, for which the tilt of the East to West arc alternates between South and North. For these cases or in general, fixed angles assigned to each cardinal point can be better. Thus, if zero degrees is North and 270 degrees is West, then the range of azimuth angles are 66 degrees to 294 degrees during Summer solstice and 114 to 246 degrees during Winter solstice. The range of azimuth angles in spring and autumn equinoxes are 90 degrees to 270 degrees. Figure 9.4 is a two dimensional plane that illustrates the sun's trajectory in the Summer and Winter solstice, as well as the Spring and Autumn equinoxes as seen in Quito.

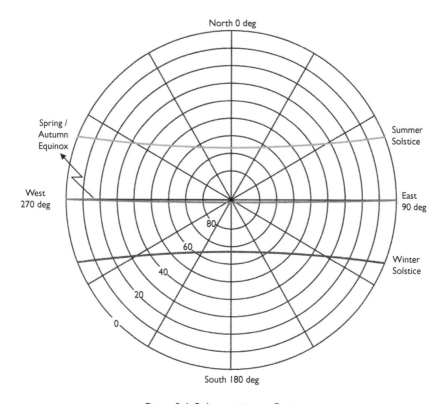

Figure 9.4 Solar position – Quito.

9.2.2 Atmospheric disturbance or clearness index

The components of the solar irradiance at ground level and components of relevance for photovoltaic applications are the following:

- Direct Normal Irradiance (DNI): Minimally disturbed irradiance traveling in a straight line from the direction of the sun. Clear days with no clouds are high in DNI. A plane normal to the sun's position would capture the DNI. This can be achieved by mounting the plane on a dual-axis tracker.
- Diffuse Sky Radiation: The irradiance that is scattered by the atmosphere. Some of this irradiance is reflected to the ground with a random incident angle. High values are typical on cloudy days or with an unclear atmosphere.
- Albedo: Irradiance reflected on the ground or other objects and captured by the plane or surface of interest.
- Global Horizontal Irradiance (GHI): This is the total of irradiance received from the sky by a surface horizontal to the ground. GHI is the sum of the DNI and the diffuse component. This is the most common measured value available for solar research and engineering.

Solar Radiation: Radiant energy (electromagnetic waves) emitted by the sun. The spectrum of solar radiation extends from high energy ultraviolet (UV, 100–280 nm) to far infrared (IR, 2500 nm), peaking in intensity at the visible band (380–80 nm). The solar radiation measures the energy of the full spectrum and the units are Joules (or Watt-hour, Wh).

Solar Irradiance or Solar Power: Radiant energy per unit time (Watts, W) incident on a surface (m²). This is a common unit used in solar engineering (W/m²).

Insolation or Irradiation: Sometimes wrongly used to refer to solar irradiance. This unit refers to solar radiation (energy) incident on a surface during a specific period of time. Typical measuring periods are day, month and year. The common unit is kWh/m² per day, month or year.

The atmosphere scatters and absorbs the extraterrestrial irradiance. Even on a clear day, approximately 25 percent of the irradiance available in the outer atmosphere does not reach the ground. Clear sky models have been developed to assess the cloudiness index or clearness index from irradiance measurements on the ground. This information is useful for applications that require DNI, such as Concentrating Photovoltaics (CPV). Otherwise, the Typical Meteorological Year datasets already capture the clearness index in the hourly values.

9.2.3   Inter-annual variability

Given the short term variation of the solar resource due to climate and environmental and atmospheric phenomena (tropical storms, changing jet streams, volcano eruptions, etc), it is necessary to collect several years of data in order to develop a long term trend of solar resource variability. Several methods have been developed to correct long term modeled data using short-term measured data and to reduce the uncertainty to +/–5 percent. Inter-annual variability of DNI is much higher (twice as much or more) than that in GHI.

Typical meteorological year (TMY) datasets are in principle, reflecting the long-term trend and thus, absorbing the inter-annual variability in the hourly values. However, it's very important to keep in mind that solar projects are financed based on a revenue model with forecasted energy production targets. These energy production targets use TMY data as the input to photovoltaic models. At the end of the year, the actual energy produced by the plant is compared to the target and the revenue generated. Therefore, it is important to understand how much less energy than the original target could be expected at the end of a year with poor solar resource. It is desired to narrow this uncertainty as much as possible in order to develop a more precise revenue model while at the same time, having the revenue model include enough tolerance to accommodate years with insolation lower than typical. A deeper knowledge of the expected inter-annual variability at the project location will reduce risk and can help to provide better financial estimates and contractual terms.

Finding the solar irradiance data applicable to the location of the project is one of the main challenges in the energy output calculations of any solar system. This is because of the generally limited or absolute lack of measured solar data at any given location around the world. Satellite technology, image processing methods, meteorological science and interpolation methods have enabled indirect measurements of solar irradiance on the earth's surface. These techniques have also enabled the

Table 9.1 Characteristics of solar resource data.

Data Type	• Ground measurements • Satellite data • Modeled data (synthetic time series)
Data Quality	• Length of time series • Data completeness • Accuracy of instruments • Data sampling rate (minute, hourly, daily, monthly, yearly) • Errors and biases
Variability	• Long term patterns • Typical year • Inter-annual variability • Uncertainty

generation of synthetic time series based on best available data for a specific location. Table 9.1 provides an overview of solar resource data characteristics.

Ideally, ground measurements would provide the most accurate representation of the solar resource patterns at that specific location. The main challenge with ground measurements is the limited number of stations available and potential data quality issues when the stations exist. Sometimes, a meteorological station is installed a priori at the location of a future PV power plant project but even in this case, the short length of the time series (often less than a year) severely limits the quality of the data set and it should not be considered as the sole source of solar resource data.

A typical meteorological year is a data set comprising hourly values of solar radiation and meteorological variables covering a full year, or 8760 hours. These datasets are assembled with long time series of measured data through a method that selects and concatenates this data into one single year. The goal is to generate a representative year that filters the extreme variations from year to year but captures the long term trend. For this reason, a TMY dataset is not necessarily a good indicator of the conditions expected for this year or any particular year. The TMY datasets are highly valuable for computer simulations.

The designer has to develop certain criteria to select the solar resource data set that is thought to best represent the location of the project. To date, there is no standard method to guide the selection of the solar resource data.

In North America, it is worth noting the work done by the National Renewable Energy Laboratory (NREL), which has developed and maintained a comprehensive solar radiation database for the continental territory of the United States as well as Hawaii, Alaska and northern Mexico[14]. Canada is not as well-mapped, although the Canadian Weather Office (Environment Canada) maintains a database of measured data from meteorological stations distributed on its territory[15]. Mexico and the remaining countries in the Americas tend to have only very localized records if at all

---

14 http://www.nrel.gov/rredc/solar_resource.html
15 http://www.weatheroffice.gc.ca/

Table 9.2 Solar resource databases.

Source	Data	
NREL	TTMY	This was NREL's first release of a TMY dataset based on data from 1952 to 1975. Due to data quality issues, the use of this data set is not recommended.
	TTMY2	The second release of a TMY dataset was completed in 1994 and is based on data collected from 1961–1990 by the National Solar Radiation Data Base (NSRDB). TMY2 data is available for 239 stations. Each station is classified based on the amount of measured data available to compile the TMY, Class A to Class C, with the former having the best quality data.
	TTMY3	TMY3 dataset is the third release of NREL's TMY and is based on the updated NSRDB data from 1991–2005. TMY3 data is based on ground and satellite measurements, including the time series used for the TMY2 dataset. This dataset contains 1,454 sites classified as Class I to Class III, with the former having the best quality data.
	TTDY	This dataset uses hourly radiance images from geostationary weather satellites, daily snow cover data and monthly averages of atmospheric water vapor, trace gases, and the amount of aerosols in the atmosphere to calculate the hourly total insolation incident on a horizontal surface. The methodology was developed by Dr. Richard Perez. The data is available on a 10 by 10 kilometer grid published on NREL's interactive web page (Solar Prospector).
NASA		This dataset is based on satellite data and published by NASA through its Surface Meteorology and Solar Energy website.
Meteonorm		Meteonorm is a computer program developed by Meteotest, a meteorology and environment company based in Switzerland. The TMY set generated by Meteonorm is based on ground measurements. This software generates a TMY for any location using an interpolation algorithm that relies on all measured data available in the vicinity of the location. This method calculates solar radiation, temperature and other meteorological parameters.
3Tier		3Tier is a private company that provides TMY datasets for any location worldwide. The dataset is based on half-hourly, satellite images from 1998 to date. The information collected is processed with proprietary algorithms and peer-reviewed methods published in scientific literature.
Environment Canada		This government agency does not generate TMY datasets. Historical and current measured solar and meteorological data from ground stations is publicly available.

existent and there are often data quality issues. The use of synthetic time series is regularly the best option for selecting data in these regions, with notable exceptions.

For energy output projections of PV power plants in the United States, it is regular practice to use NREL's database as the primary source of datasets. Despite the comprehensiveness of this database, the designer should still verify the quality and applicability of the dataset to the project location. Other sources of solar resource datasets are indicated in Table 9.2.

9.3 PROJECT REQUIREMENTS

To follow an efficient design process it is necessary to clearly understand the needs and limitations of the project. For utility scale projects, the most significant requirements defined by the developer, owner, utility or grid operator are specific DC or AC capacity targets, specific production profiles (maximizing production at defined seasons or times of the day) and interconnection characteristics. The location itself also imposes certain limits or challenges on the system due to soil conditions, topography, hydrology, environmental loads (wind, snow, seismic), surface area, environmental impacts and infrastructure available.

The DC capacity of a project is generally defined in terms of the nominal power of the photovoltaic module[16] and expressed in units of Watts-peak (W_p)[17]. The nominal power is measured at the factory under controlled conditions referred to as Standard Test Conditions (STC) and clearly defined in international standards[18]. The DC capacity expressed as W_p or STC is equivalent. There are also other standards to define the nominal power of PV modules such as Photovoltaics for Utility Scale Applications (PVUSA) Test Conditions (collectively known as PTC), Nominal Operating Cell Temperature (NOCT) and others[19]. Using any of these standards, the DC capacity of a PV plant can be defined without ambiguity. However, the AC capacity can be defined in several manners, and there is no standard or industry agreement to date regarding a consistent AC capacity definition. In practical terms, the parameter of relevance is the maximum AC capacity at the point of interconnection, and while the DC capacity is key for design purposes it is not important for interconnection. Some of the most common AC capacity definitions found in the industry are the following:

1) *Inverter Capacity:* The AC capacity of the plant is based on the sum of the nameplate capacity (nominal AC power) of the inverters in the power plant. Given that what is important is the AC capacity at the point of interconnection, the nameplate capacity is often used only as a starting point. This capacity is derated by considering all the losses that occur from the inverter output to the point of interconnection. These components are generally a medium voltage transformer, wiring and breakers. The derating is typically applied as a percentage loss.

    The advantage of this method is that it provides a straightforward definition and fast calculation of the AC capacity of a system. However, the actual AC capacity achieved by a photovoltaic system depends on its DC capacity and the amount of irradiance incident on the modules. Therefore, the maximum AC capacity based on the nominal capacity of the inverter will not provide an AC profile of the photovoltaic system throughout the year, but an indication of the maximum AC power that can occur some days during the year.

---

16 The sum of the nominal power of each photovoltaic module used in the power plant.
17 And units multiples, e.g. kW_p, MW_p.
18 Standard Test Conditions are 1000 W/m^2 with a spectral distribution of AM1.5 as in ASTM standard G173-03 (2008), and 25°C cell temperature.
19 PTC are 1000 W/m^2 with a spectral distribution of AM1.5 (ASTM G173-03) and ambient temperature of 20°C. NOCT are 800 W/m^2 with a spectral distribution of AM1.5 (ASTM G173-03) and ambient temperature of 20°C.

2) *CEC-AC Capacity:* This rating is required by the California Solar Initiative program and is based on the DC capacity of the system and the efficiency of the inverter. This method makes use of the nominal power of the modules measured or estimated under PVUSA Test Conditions rather than under STC. The main difference between STC and PTC conditions is the cell temperature. While the nominal power at STC refers to a cell temperature of 25°C, the nominal power at PTC refers to an ambient temperature of 20°C, which translates to an approximate cell temperature of 50°C (depending on module technology and the specific model). As a consequence, the nominal power of the module rated at PTC will be lower than that at STC due to the temperature effect[20]. The goal of PTC rating is to reflect field conditions, which are better captured defining an ambient temperature rather than a cell temperature. The CEC-AC capacity of a system is calculated as the product of the nominal DC power (PTC) times the inverter efficiency[21]. This method effectively correlates the DC capacity of the system to the AC capacity output following specific ratings for the module and the inverter[22]. Because of the use of PTC rating of the modules and a weighted efficiency, this rating provides a closer AC rating of the system than the inverter capacity method.

3) *AC Capacity at Operating Conditions:* This method is a step further than the CEC-AC method and requires the use of a program to calculate detailed module temperature and AC energy generated at the inverter output. The program can be custom built or the designer can rely on commercially available software for PV applications. Using hourly irradiance data, it is possible to estimate the cell temperature reached by the module depending on the angle of incidence and the heat absorption of the cell. The heat absorption can be calculated based on cell efficiency, cell material, structure of the module, irradiance reaching the cell and heat removed by air flow (which depends on the mounting type and wind speed). However, wind speed around the modules is often not known. The cell temperature calculations can be simplified and relatively accurate estimates can be obtained. It is important to use hourly data due to the variability of irradiance throughout the day.

Based on the results of the program over a whole year of typical irradiance, it is necessary to implement some type of statistical analysis in order to determine the maximum AC capacity that the system is likely to achieve within certain limits of confidence. Because the data has hourly resolution, it is also possible to estimate AC capacities on a monthly basis, which can be very useful as the solar patterns may change significantly throughout the year, thus changing the maximum expected output of the system as well. As mentioned earlier, no standard has been developed to date. Thus any reasonable approach can be proposed and accepted by the developer, owner, investor or other entities vested in the project.

In addition, detailed loss calculations for the wire, transformer and other components can also be included.

---

20 Higher operating temperatures of the cell reduce the photovoltaic conversion efficiency.
21 The CEC inverter efficiency (refer to section 9.4.2).
22 The California Solar Initiative program (CSI) maintains a database of PTC ratings of modules and CEC efficiencies of inverters (http://www.gosolarcalifornia.org/links/equipment_links.php).

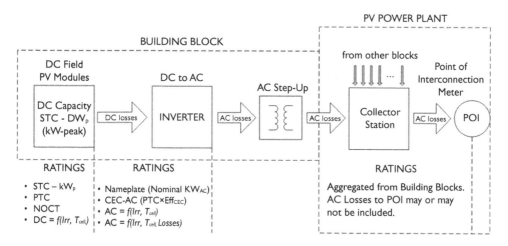

Figure 9.5 Capacity ratings.

Figure 9.5 illustrates the most common AC ratings, starting with the most typical case (nameplate capacity). The last case (capacity at operating conditions) is becoming more standard for large scale installations. The AC capacity target and definition used by the developer will enable the designer to best match the capacity of the DC field to that AC target.

At the same time, another important project requirement that the designer must be aware of is the expected profile of the generated energy. Large scale photovoltaic systems, particularly rooftop projects, are normally designed to maximize the energy output on an annual basis because this approach is typically the best option for maximum revenue. However, in some instances, most commonly in utility scale projects, the best revenue option can be the use of the photovoltaic system as a power generator designed to meet specific peak demands at specific times of the day (or the year), in which case the project may not generate the maximum energy possible on an annual basis. In either case, the designer has to select the best orientation and tilt to maximize the energy production of the systems that matches the specified energy output profile. The selection of the optimal tilt and orientation is best accomplished with the use of software that can generate multiple cases in a short period of time, taking into account the variation of the solar resource on very short time scales throughout the year (hourly or sub-hourly data points). The designer has to negotiate the results of the optimal tilt and orientation with a practical mounting structure (including foundations) that is commercially available, suited to the geology and topography of the site, meets the local wind requirements, and is a cost effective solution. The effectiveness of the mounting structure is typically assessed in terms of its capital, operation and maintenance costs, compared to the revenue it contributes to generate. Thus, the best analytical tilt and orientation is sometimes not the most practical and cost effective. In addition, the actual surface area available for the project can limit the arrangement options of the mounting structures, which may reduce capacity or increase the inter-row shading, in both cases reducing the energy production.

These are some reasons why it is important to understand the site conditions and mounting options available as early as possible in the development of the project. It is recommended that geotechnical studies have a high priority in the timeline of project development.

It is important to note that systems using dual-axis trackers follow the sun at all times so their tilt and orientation are always optimal[23]. However, it is still necessary to consider foundation options, local wind requirements and cost analysis before selecting this option.

The design team has to follow an optimization process in order to match the AC capacity requirements, production profile and selection of the mounting structure. Figure 9.1 illustrates this iterative loop. In addition, the specific module technology will also impact the optimization process. The module efficiency will define the surface area needed to achieve the required AC capacity while the module's framing characteristics will define the mounting structure requirements to hold the module in place. The interconnection requirements will define the type of inverter necessary for the specific project. From this perspective, grid-tied inverters can be classified in two main categories: anti-islanding enabled and low voltage and frequency ride through (LVFRT) with reactive power supply capabilities. These features will be discussed later in this Chapter.

9.3.1 Photovoltaic module

The photovoltaic module and the inverter are the two fundamental components of a photovoltaic system; the former enables the power generation and the latter enables the conversion to standard AC power.

The photovoltaic modules transform the sunlight into electricity. Current photovoltaic conversion technologies are based on semiconductor materials, with crystalline silicon (cSi) being the dominant technology with a total global market share of approximately 80 percent at the end of 2011[24]. Thin film technologies, which have become important players in the past few years, complete the remaining market share. Cadmium Telluride (CdTe) is the dominant thin film module technology, but Copper-Indium-Gallium-(di) Selenide (CIGS) modules are gaining momentum. Another commercial thin film module technology is based on amorphous silicon (aSi). The most efficient cells are the multi-junction cells, which are used in concentrating photovoltaic (CPV) systems. CPV systems have been implemented in large commercial projects since 2010, but as of the end of 2011, they still represented less than 0.2 percent of total market share.

Each of these photovoltaic technologies presents different operational characteristics, which may offer a performance advantage under certain conditions. Table 9.3 summarizes the most salient performance differences between the four commercially available technologies to date as of 2012. Keep in mind that the price of the module represents about 50 percent of total project costs in large scale systems and thus, the performance advantages are sometimes compromised to favor the overall economics of the project.

---

23 Concentrating Photovoltaic (CPV) systems inherently use dual-axis trackers.
24 Photon magazine, March 2012.

Table 9.3 Module technology – summary of features.

Module technology	Performance notes
Mono-crystalline silicon (mSi) Poly-crystalline silicon (pSi)	Crystalline silicon modules (mono and poly) have the longest commercial history of all photovoltaic technologies dating back to the late 1970s (mono-crystalline modules). This experience has helped the industry to achieve a high degree of product reliability and market confidence. Mono-crystalline silicon modules present a better performance at higher ambient temperatures and have a slightly higher efficiency than poly-crystalline silicon modules. Typical power temperature coefficients for mSi modules are between –0.39 and –0.44 (%/°C) while pSi modules are between –0.44 and –0.47 (%/°C). This coefficient defines the power degradation for every degree Celsius increment in cell temperature. Typical module efficiencies are around 16.5 percent for mSi and 14.5 for pSi. On the other hand, mono-crystalline modules are more expensive than poly-crystalline modules, and for large scale projects, the latter are often selected for this reason.
Cadmium Telluride (CdTe)	Commercial modules based on this semiconductor alloy were first used in large systems around 2004 but since then, this type of module has been used in several of the largest photovoltaic installations in the world. CdTe modules have a better temperature response than mono-crystalline modules but a lower efficiency. The typical power temperature coefficient is –0.25%/°C while the efficiency of these modules is between 10 and 12 percent. This technology also enables a better response of the module at low irradiance conditions. This means that the efficiency curve of CdTe modules is less steep at low irradiance conditions than the efficiency curve of cSi modules under the same conditions. The thin film manufacturing processes enable low production costs, and hence the price of these modules are the lowest in the photovoltaic module market as of 2011.
Copper-Indium-Gallium-Selenide (CIGS)	The commercial history of these modules dates to 2010 making this technology one of the most recent additions to the photovoltaic module market. The market confidence of these modules is building up and some large scale installations are being developed with this module. The temperature response of these modules is in between the crystalline silicon and CdTe modules, around –31%/°C, while efficiency is comparable to CdTe modules, around 12 percent. These modules exhibit a "Light Soaking" effect, in which the output power can be as much as 10 percent higher than the nominal output power after the module is exposed to sunlight for the first time. This higher output decreases to the nominal power output in the course of about three months. As of Q2-2012, the price of these modules tends to be slightly lower than crystalline modules but, due to the thin film manufacturing processes, it is expected that prices will drop rapidly when high volume manufacturing and market demand ramps up.
Multi-junction cells (For CPV applications)	These cells present the highest efficiency and the best temperature response of all commercially available photovoltaic technologies, as well as the highest cost. Typical efficiencies as of 2012 are in the range of 37 to 40 percent [references] while typical temperature coefficients are between –0.15%/°C and –0.17%/°C. These cells have been used in space applications since the late 1990s and are used in CPV systems. The first large commercial projects were installed at the end of 2009. Due to the inherent use of dual-axis tracking (in order to collect the DNI), higher efficiency and better temperature response, CPV systems can achieve a greater energy production than any other technology on a per Watt basis.

There are also variations in quality and performance from vendor to vendor, particularly in the crystalline silicon market, which has the largest diversity, with over 200 manufacturers worldwide. This situation has given rise to so-called "Tier 1", "Tier 2" and "Tier 3" type cSi modules, with Tier 1 modules assumed to have a high manufacturing quality, high field performance and long-term reliability. These categories are not clearly defined but are often used by investors and developers to differentiate and rank cSi manufacturers. Some of the general qualities assumed to assign ranks are, for example, the manufacturing capacity (volume in MW/year) of the company, global sales to date, financial health of the company, number and type of projects built with the company's product(s), field history of the product and company's and product's reputation within the solar industry.

The diversity of vendors for thin film manufacturers is much narrower, with one single company sharing well over 50 percent of the global thin film module market with few other important players sharing the rest[25].

For the most part, the selection of the module technology for the development of a photovoltaic project depends on two factors: module cost range set by the economics of the project and specific site conditions. However, the financial analysis should take into account the different qualities of the photovoltaic technologies as well as the advantages and disadvantages of them in the economics of the project. For instance, the site conditions can highlight the technology that best matches those conditions. For example, a rooftop system will maximize the power installed in the limited surface area of the building by using the most efficient mono-crystalline modules. In arid locations, the use of thin film modules or CPV systems may be advisable due to their better temperature response and adequate DNI conditions for CPV. In general, the major disadvantage of thin film modules is that the lower efficiency requires more surface area to reach the same nominal power as any other photovoltaic technology. From this perspective, CPV systems are more compact but also likely to be more expensive to install[26].

Poly-crystalline modules and Cadmium Telluride modules are the dominant technologies in large scale systems. Mono-crystalline modules are sometimes used, but in general, the higher cost of these modules can make the project economics not as competitive as using pSi or CdTe modules. Large scale CPV systems have also taken off since 2010, totaling approximately 360 MW in operation or development as of Q2-2012. CIGS and amorphous silicon modules may become an important thin film player in the implementation of these large systems in the near future. Poly-crystalline and mono-crystalline modules dominate the rooftop and residential market around the world, with some rooftop and residential applications being implemented with thin film modules, particularly in Europe.

25 As of 2012, First Solar is the leading thin film manufacturer of CdTe modules. Other thin film manufacturers include Abound Solar, Solar Frontier, Ascent, Q-Cells, Miasole, Kaneka, Nanosolar and many others.

26 The capital costs are only one aspect of the total costs of a project. The amount of electricity generated over the life time of the project is another critical aspect. Due to the use of dual-axis tracking, CPV systems can generate more electricity if the DNI on site is relatively high and constant throughout the day. The other PV technologies do not require high DNI to operate efficiently.

9.3.2   Photovoltaic inverter

The Inverter is the power conversion unit that transforms the DC power generated by the modules into AC power. Modern inverters include advanced electronics that enable fine control of the conversion process as well as a wide range of other features such as continuous monitoring of internal and grid parameters. In the case of large capacity inverters, the electronics become more sophisticated and also include data acquisition, telemetry, telecommunications and other supervisory and control functions.

Photovoltaic systems can be installed as stand-alone, grid-tied and hybrid systems, each one making use of inverters that have been designed to operate for that application and are not interchangeable (with the exception of the hybrid).

The stand-alone systems use the modules as their only source of energy. This energy can be stored (in batteries for example) or used as it is generated. The hybrid type of solar inverter is capable of operating in stand-alone mode as well as grid-tied mode. The applications and photovoltaic systems designed with these two types of inverters are not discussed in this book.

Grid-tied systems make use of an inverter that delivers all the energy generated by the modules to the existing electrical infrastructure. These inverters "plug-in" to the grid. The flow of energy is unidirectional, from the modules to the inverter to the grid. Grid-tied inverters represent the majority of the inverter market and they are available in power ratings as low as 700 W to 1,000,000 W (1 MW) and a few even larger. There is a growing interest in developing energy storage solutions for grid-tied systems to address the intermittency of the solar resource. The integration of energy storage on a large scale will reshape the inverter characteristics, and hybrid inverters may become more prevalent.

The inverter is, in essence, an electronic circuit that converts the DC power into AC power through switching[27]. The main components of the converter are power electronic devices that are able to change from an ON to an OFF state at a very high rate; 3 to 8 kHz is common[28]. These devices, typically Insulated Gate Bipolar Transistors (IGBT), are also capable of handling power in the order of tens to hundreds of kilowatts per unit. IGBTs are a relatively recent achievement in mass manufactured power electronics technology[29]. Improvements in other semiconductor technologies, most notably silicon carbide (SiC) devices, with a better performance response than silicon semiconductors, will also be reflected in newer power electronic devices used in inverters[30]. Other important components in the inverter are high voltage and high current capacitors and inductors. Figure 9.7 illustrates a typical converter circuit.

27 Typical configurations are full and half-bridge circuits.
28 New topologies with switching frequencies over 20 kHz are also being developed.
29 Metal oxide semiconductor field-effect transistors (MOSFET) where the immediate IGBTs predecessors in high power inverters. Modern IGBTs have overcome some of the early disadvantages they had when compared to MOSFETs, such as lower commutation rate, important losses at low current levels and latch-up. IGBTs are capable to operate at higher voltages than MOSFETs and are the dominant power electronic technology for utility scale inverters. MOSFETs are still widely used in low power inverters.
30 These improvements can include significantly higher breakdown voltage, lower switching losses, higher thermal conductivity and higher temperature operation capability.

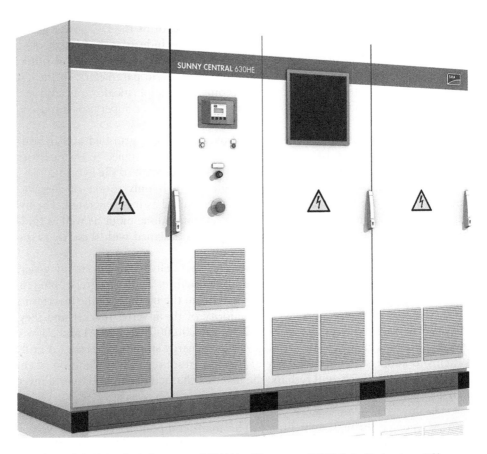

Figure 9.6 Utility Scale Inverter – 600 kWac (Courtesy of SMA Solar Technology AG).

The typical conversion efficiency curve of these inverters is relatively flat, with DC power input higher than 20 percent of total capacity, dropping dramatically with power input less than 10 percent of the total. The highest efficiencies tend to be between 25 and 75 percent of total power input capacity. The power conversion efficiency depends on converter design, input voltage levels and operating temperature of the inverter. The latter is a function of input power and ambient temperature and is mitigated by cooling mechanisms. Forced air cooling (blowers) and liquid cooling are used in commercial inverters. If a certain temperature threshold is reached, the inverter will curtail the power conversion capacity to avoid overheating. The power output performance as a function of the ambient temperature is not typically provided in the data sheet and has to be requested from the manufacturer.

The efficiency is typically reported as a single value that does not accurately reflect the variation of efficiency at different power inputs. In North America, the California Energy Commission (CEC) has set a standard, based on the Sandia Inverter Performance Test Protocol (developed by Sandia Laboratories), that has become widely

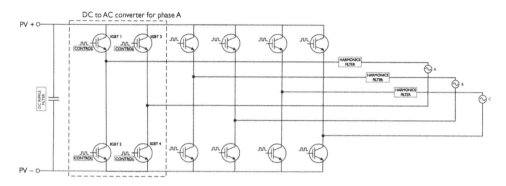

Figure 9.7 DC to AC Converter (full-bridge).

adopted by the industry. The CEC efficiency is a weighted efficiency that is calculated using data measured at various power inputs. Typical CEC rated inverter efficiencies for large capacity inverters are around 97.5 percent. The efficiency curves of the inverters as a function of the DC power input and DC voltage input are not typically provided in the data sheet and have to be requested from the manufacturer.

The core of the inverter operations and control functions reside in the electronic controller of the unit. This electronic controller is a microprocessor-based, special purpose computer that enables the overall function of the inverter and also performs a number of safety operations. The main inverter functions can be divided into four main categories:

a) Grid synchronization and control: The controller monitors the voltage and phase of the grid, which is used as the reference signal to define the ON and OFF cycles of the IGBTs bridges. The ON and OFF cycles determine the qualities of the DC to AC conversion processes. The DC to AC stage compares its output signal to the reference grid signal to match phase.

b) Photovoltaic field management: Inverters manage the MPP of the DC field connected to it by continuously scanning the system voltage and setting the Maximum Power Point (MPP) based on this information. The MPP tracking algorithm must be able to adapt to transient changes in irradiance (due to fast moving clouds or other atmospheric factors), as well as to enable stable operation under low irradiance conditions. This is the most important MPP characteristic, but this information is normally not published by inverter manufacturers and has to be requested.

c) Safety functions: Inverters must comply with or exceed personal safety standards developed by internationally recognized organizations. The controller permanently monitors the DC field to detect ground faults or other failures, generate alarms and disconnect. The controller also monitors the grid to avoid islanding, to detect AC anomalies or failures and to automatically disconnect.

d) Data acquisition and communications: Modern inverters include an integrated data acquisition module to monitor and collect virtually every single parameter measured by the controller for performance and protection. Considering the capabilities of embedded systems, most modern inverters can also support Supervisory Control and Data Acquisition (SCADA) systems. Depending on the application, full SCADA may be enabled or limited to data acquisition for monitoring with no control capabilities.

The above characteristics are common to all inverters. However, there are two different types of interconnection requirements depending on the size and application of the photovoltaic system:

- Systems less than 10 MW interconnected to the distribution system, not under the control of the local utility and regulated by Electrical Codes guidelines for public installations[31]. This includes all rooftop systems and several ground mounted systems that may exceed 10 MW.

  In these systems, the inverter must stop operation if there are voltage or frequency disturbances on the grid. The tolerances for over/under voltage and frequency shift are set very narrow in these inverters in order to avoid an islanding situation. For photovoltaic inverters, islanding is defined as a condition in which a portion of a utility circuit is energized solely by one or more local photovoltaic inverters connected to that circuit. This is a hazardous condition, as utility workers or other users may wrongly assume that the circuit is de-energized. In addition, equipment damage may also occur. For this reason, photovoltaic inverters must be able to continuously monitor the grid voltage and frequency, and if a disturbance beyond the tolerance values is detected, the equipment should automatically power off and disconnect from the grid. The standard IEEE 1547.1-2005 section 5.7 defines a test to verify that a photovoltaic inverter stops supplying power to the AC circuit it is connected to. The results of this test determine the time it takes for the inverter to detect the islanding condition and stop supplying power. The maximum tripping time defined by standard IEEE 1547.1-2005 is two seconds, depending on the magnitude of the excursion from nominal. In more detail, Table 9.4 shows the voltage clearing times and Table 9.5 shows the frequency clearing times from IEEE 1547 for generation sources greater than 30 kW.

  The test defined by IEEE 1547.1-2005 is internationally recognized and most inverters for the North American market reference this standard for testing and certification. The standard UL (Underwriters Laboratories) 1741–2010 edition has been harmonized with the anti-islanding requirements stated in standard IEEE 1547.1-2005.

  The International Electro-technical Commission (IEC) has also developed a standard to specifically provide a test procedure to evaluate the performance of islanding prevention measures of grid-interactive photovoltaic inverters. The IEC standard IEC 62116-2008 is also internationally recognized and widely used in Europe. In general, the requirements of both standards, IEEE

---

31 The National Electrical Code and the Canadian Electrical Code in the United States and Canada, respectively, which may also include specific requirements requested by the Local Authority Having Jurisdiction.

Table 9.4 Clearing times following abnormal voltages.

Voltage range (% of nominal)	Clearing time (cycles)	Clearing time (sec)
50%	10	0.16
50% to 88%	120	2.0
88% to 110%	Normal	Normal
110% to 120%	60	1.0
>120%	10	0.16

Table 9.5 Clearing times following abnormal frequencies.

Frequency range (Hz)	Clearing time (cycles)	Clearing time (sec)
57 Hz	10	0.16
57 Hz to 59.8 Hz	10 to 18	0.16 to 300
59.8 Hz to 60.5 Hz	Normal	Normal
>60.5 Hz	10	0.16
57 Hz	10	0.16

1547.1-2005 and IEC 62116-2008 are compatible[32]. The maximum tripping time defined by standard IEC 62116-2008 is also two seconds with voltage and frequency intervals compatible with Table 9.4 and Table 9.5.

• Systems over 10 MW connected to the transmission system (directly for very large systems or through a distribution circuit for smaller systems) and under the control of an electrical utility. Systems under the control of an electrical utility and with access restricted to authorized personnel are not considered public installations and typical Electrical Code regulations do not apply. These systems are commonly referred to as "behind-the-fence" installations.

With the advent of multi-megawatt photovoltaic systems and higher penetration levels of photovoltaic and wind projects, there has been a growing concern regarding potential disturbances to the transmission systems. In Europe, the German Association of Energy and Water Industries (BDEW by its German acronym) proposed a set of guidelines for photovoltaic power plants interconnecting to the medium voltage network. These guidelines were introduced in 2008 and all photovoltaic plants in Germany under this category are required to fully comply with them as of April 2011. The main goal of the guidelines is to reduce potential disturbances to the transmission system and maintain grid stability. To achieve this goal, the photovoltaic power plants are expected to have a more traditional generator performance,

---

32 The requirements for passing the anti-islanding test in IEC 62116-2008 includes more test cases than IEEE 1547.1-2005 but the test circuit and the conditions for confirming island detection do not have a significant deviation from IEEE 1547.1–2005.

although the lack of energy storage to compensate for the intermittency inherent in photovoltaic systems has not been addressed. There is a technology and cost barrier to successfully integrating energy storage in large scale power plants for commercial operation to date. Because of these reasons, the BDEW and other guidelines do not address this issue. The BDEW guidelines have been used in other European countries as a reference to develop similar standards, or are being used as the effective standard.

In the United States and Canada, utilities and grid operators often require performance features that are similar to the BDEW guidelines. No standard has been adopted in North America, and as of mid-2012 the characteristics of each medium voltage interconnection are managed on an individual basis by the utility/grid operator and the developer. The California Independent System Operator (CAISO) proposed a defined set of interconnection characteristics for photovoltaic projects. The modifications would apply to solar facilities (asynchronous generators) that are directly interconnected with the CAISO transmission system and are 20 MW or larger. Systems less than 20 MW may also be requested to meet these requirements. The proposed changes would require these solar facilities to present the same operational characteristics as conventional (synchronous) generators. The specific interconnection requirements that solar facilities would be required to incorporate are the following:

a)   Low voltage and frequency ride through (LVFRT)
b)   Power factor regulation (reactive power supply capabilities to maintain PF within at least 0.95 leading / 0.95 lagging)
c)   Voltage regulation
d)   Generator power management (remote control to dispatch and curtail generation)

The proposed changes are contained in document ER10-1706 filed by CAISO on July 2, 2010 and were rejected by the Federal Energy Regulatory Commission (FERC) at the end of 2011, on the grounds that baseline reactive power requirements should be justified by a specific interconnection study. There is an ongoing decision-making process among regulatory agencies, solar energy organizations and utility/grid operators regarding the form and application of interconnection requirements for large scale photovoltaic projects[33].

While the set of interconnection characteristics proposed by CAISO was not set as a standard, any of the four interconnection characteristics identified above have been, and can be required in future large scale solar projects on an individual project basis[34].

---

33 See for example, Sandia's report "Reactive Power Interconnection Requirements for PV and Wind Plants – Recommendations to NERC" February 2012 as well as the opinions of the Solar Energy Industries Association SEIA); Large Scale Solar Association (LSA), North American Electric Reliability Corporation (NERC); Federal Energy Regulatory Commission (FERC).

34 Each particular project dictates the specific performance parameters applicable to any of the four grid support requirements.

The implementation of LVFRT and remote power management in particular has significant performance implications for inverters:

a) Anti-islanding performance: LVFRT requires avoiding disconnection of the unit if disturbances on the grid signal occur for a defined time period (up to three seconds), as opposed to anti-islanding, which requires immediate disconnection. The objective is for the power plant to continue supplying power while the grid recovers, rather than to disconnect a significant amount of power, which would further destabilize the network.

b) Communications and SCADA capabilities: The inverter must have full SCADA capabilities in order to dynamically control set points that can change the response of the inverter at all times, such as reactive power supply (changes in PF), ramp rates, curtailment (power reduction) and over/under voltage and frequency points. The SCADA has to be able to communicate with the electrical utility's SCADA protocols and systems.

Current commercial inverters are able to implement the above Grid Support and Remote Power Management without a significant change in the inverter's architecture. As the operation characteristics of the inverter are controlled by the algorithms embedded in the electronic controller, a change in the inverter operation requires, for the most part, changes in the control software. Some minor hardware implementations are necessary, such as some energy storage to continue operation while the grid recovers, a few additional sensors and some additional signal conditioning circuitry.

In the future, it is possible that smaller systems will also be required to meet similar requirements, assuming high penetration levels of photovoltaic systems together with large scale implementation of smart-grid networks.

As described above, the inverter is not only the energy conversion unit but the main control center of the photovoltaic system. While the basic power electronic circuit and signal conditioning circuits are in essence the same across all manufacturers, the computational power of modern microprocessors has enabled the inclusion of complex algorithms and multi-function capabilities to this equipment. Therefore, the control mechanisms embedded in the electronic controller are more unique and specific to each manufacturer. As a consequence, it is recommended that the designer requests measured performance data from the potential vendors, particularly regarding the efficiency conversion curve, MPP tracking under unstable and low irradiance conditions, power derating due to temperature and actual reactive power supply capabilities. Nominal data sheets will look very similar across all inverter manufacturers but measured performance data will provide more insight into the actual response of the inverter over a wider range of conditions.

Other important differences among inverter manufacturers are the reliability program followed by the manufacturer and the reliability tests performed on the equipment and components in order to estimate mean time between failures and repairs (MTBF, MTBR), as well as expected lifetime of the product. Although basic hardware design may not change from manufacturer to manufacturer, differences in the quality of the components and quality in manufacturing exist and will certainly affect the long term performance and availability of the inverter. Therefore, it is recommended that the designer takes the time to request and research the measured performance and reliability data of the equipment.

Other performance and safety tests are covered by international standards that the product should comply with. The test certificate has to be applicable to the specific inverter model selected and the test has to be performed by a recognized testing laboratory. There are a number of standards applicable to inverters that have been drafted by different organizations (e.g. IEEE, UL, IEC, EN), although most of them are equivalent. Most IEC standards for inverters are harmonized with the European Standards (EN) and in many instances they can be more comprehensive than the IEEE and UL standard, although the latter are more recognized in the United States and Canada[35]. Table 9.6 provides a list of the minimal standards that new commercial inverters are expected to meet.

The input to the inverter is generally referred to as the DC field and includes all the equipment, parts and materials that support and connect the modules to the main input terminals of the inverter.

9.3.3 Module mounting structure

The mounting structure provides a permanent mechanical support for the modules of the photovoltaic system. Although this component is not as technologically complex as the inverter or the module, a well-engineered mounting structure is critical to ensure the specified performance and long term reliability of the whole photovoltaic system. The basic design of mounting structures for large scale photovoltaics is defined by the two main market segments for these types of projects, which are very large rooftops and open spaces at ground level. The most accessible and cost-effective roofs on which to install photovoltaic systems are flat rooftops of large buildings such as warehouses. Tilted roofs can also accommodate relatively large systems although in general, the space available on these roofs is not as significant as flat rooftops and the installation is often more expensive. Ground mounted systems are hosts to so-called utility scale projects, which have grown steadily since the early 2000s in Europe and since 2007 in the United States.

Perhaps due to the relative simplicity of the mounting structures as well as their relative cost, this component was often not as scrutinized as the module or the inverter in earlier large commercial systems. However, with the development of a significant market for large rooftop and ground mounted systems in Europe and in North America, the importance of selecting a reliable and durable mounting system should not be underestimated by the designer. In addition, with the steady decline in the price of modules and inverters, the mounting structure has become a very significant expense of total system costs, typically representing the second largest expense, ahead of even the inverters. The most expensive component category remains with the modules.

To date, most mounting structures for these two market segments are made of metal, largely aluminum and hot-dip galvanized steel, although a few vendors are offering structures made of ultraviolet (UV) resistant, structural plastics for the rooftop market.[36] In general terms, the main components of a mounting system are described in Table 9.7:

---

35 International Electrotechnical Commission. European organization.
36 Ultraviolet light (UV).

Table 9.6 Typical inverter standards.

Standard	Description
IEEE 1547 (2003)	Series of standards: 1547 / 1547.1 / 1547.2 / 1547.3 / 1547.4. This standard focuses on the technical specifications for the testing and for the interconnection itself that a photovoltaic inverter must meet prior to the connection to the utility; at the instant of connection; and for proper disconnection if abnormal conditions are present. The standard also provides requirements relevant to the performance, operation, testing, safety considerations, and maintenance of the interconnection. It includes general requirements, responses to abnormal conditions such as islanding, power quality and test specifications and requirements for design, production, installation evaluation, commissioning, and periodic testing of the interconnection.
UL 1741 (2010)	This standard defines minimum requirements for the design and manufacture of power inverters, converters and charge controllers, and interconnection system equipment for use with distributed energy resources. This standard has been harmonized with IEEE 1547 requirements for interconnection.
IEC 62116 (2008)	This standard describes a guideline for testing the performance of automatic islanding prevention measures installed in or with single or multi-phase utility interactive PV inverters connected to the utility grid. The test procedure and criteria described are minimum requirements that allow repeatability. Additional requirements or more stringent criteria may be specified if demonstrable risk is shown. Inverters and other devices meeting the requirements of this standard are considered non-islanding.
IEC 62109-1 (2010)	This is a comprehensive standard, newly-released (2010) and specifically tailored to power conversion equipment for use in photovoltaic systems. This standard defines the minimum requirements for the design and manufacture of power conversion equipment for protection against electric shock, energy, fire, mechanics and other hazards. This standard provides general requirements applicable to all types of power conversion equipment, and Part 2 (released on June, 2011) provides specific requirements for inverters. This standard covers equipment connected to photovoltaic systems which does not exceed a maximum source circuit voltage of 1,500 VDC. The equipment may also be connected to systems not exceeding 1,000 VAC at the AC mains circuits, non-main AC load circuits, and to other DC source or load circuits such as batteries.
IEC 62109-2 (2011)	This Part 2 of IEC 62109 covers the particular safety requirements relevant to grid-interactive and stand-alone inverters, wherein the inverter is intended for use in photovoltaic power systems. This equipment has potentially hazardous input sources and output circuits, internal components, and features and functions, which demand different requirements for safety than those given in Part 1.
IEEE-519 (1993)	This standard defines the requirements for harmonic control in electrical power systems.
IEC or EN 61000-3-4	The family of standards under EN 61000 refers to electromagnetic compatibility (EMC) covering emission of harmonic currents and immunity.
IEC 60204-1	This standard defines minimal requirements and recommendations for electrical and electronic equipment and systems in order to promote safety of persons and property, consistency of safety control mechanisms and ease of equipment maintenance. This standard covers all the electrical systems of the machine through the point of connection. This standard is applicable to electrical equipment operating with nominal supply voltages not exceeding 1000 VAC or 1500 VDC and with nominal supply frequencies not exceeding 200 Hz.
IEC 62103 (EN 50178) Current standard is IEC 62103	This standard defines the minimum requirements for the design and manufacture of electrical equipment connected to voltage levels not exceeding 1000 VAC or 1500 VDC. This standard is used as a reference for the electrical equipment used to assemble the inverter; such as breakers, switches and disconnects, surge arresters, and other equipment for protection against electric shock.

Table 9.7 Mounting structure – main components.

Component	Description
Foundations or Anchors	These components serve to support and anchor the frame of the mounting system to the surface where the photovoltaic system is going to be installed.
Structural Frame	This assembly of parts provides the main structural support to which the modules are attached, either directly to the frame components, or through rails specifically designed to secure modules.
Rails	The rails are designed to permanently secure the modules to the structural frame. The actual fastening of the modules is achieved directly through the mounting holes in the module's frame, by clamps or other methods that hold the module from the edges (which is the non-active surface of the module). The rails are the base for these engineered clamps that ensure the module is permanently fastened to the structural frame. Depending on the mounting structure, the fastening method can be directly integrated into the structural frame, which eliminates the need for a rail subsystem.
Clamps	The clamps are the mechanical interface between the module (framed or glass-to-glass) and the rails or structural frame. The clamps fasten the modules to the rail or structural frame by applying a compression force from the edges of the module to the rail or structural frame.
Electrical Connections	The mounting structure must have provisions for grounding or electrical isolation between the frame and the metal structure. Some products also offer wire management features.
Electronic Controller	This component is only used in trackers, which require a motion control system to regulate the rotation of the tracker relative to the sun's position throughout each day of the year. This component is a micro-processor-based electronic board with an embedded algorithm that calculates the sun's position relative to the geographical location of the tracker system. The results of these calculations drive a motor or actuator that provides the mechanical torque to the tracker system. The algorithm of most trackers also includes backtracking performance. Often, the electronic controller also performs some monitoring and safety functions and may also have data acquisition and communications capabilities.

From the design perspective, the mounting structure defines two fundamental aspects of the photovoltaic system:

1) The final layout of the system
2) The degree to which the modules are exposed to the sun throughout the day and the year

Although the final layout of the system is primarily determined by the actual topography and contours of the site where the system will be installed, the layout is strongly influenced by the assembly and construction characteristics of the mounting structure selected. Different mounting structures offered in the market place have specific features to adapt to the site's conditions, which will also impact the degree of complexity required to construct the photovoltaic system on site.

The tilt and orientation of the mounting structure will determine the amount of sun received by the module throughout the year, which will impact the overall

Table 9.8 Types of mounting structures.

Application	Mounting type	Typical foundations	Typical tilts
Tilted Roof	Rails	No foundation system. Penetrating anchor points or special attachment methods.	Defined by the roof tilt.
Flat roof	Fixed tilt	No foundation system. Penetrating anchor points and ballasts.	5 to 15 degrees most common in steps of 5 degrees. Flat and over 15 degrees also available.
Ground Mounted	Carport	Concrete piers	0 to 15 degrees. High degree of customization.
	Fixed tilt[A]	Driven piles are the most common. Concrete piers, helical or screw piles and concrete ballasts are also used.	15 to 25 degrees most common. 30 degrees and higher also available. Tilt is customized.
	Horizontal, single-axis tracker (SAT)[B]	Same as above. Concrete ballasts are seldom used.	0 degrees (horizontal) with East to West movement of +/–50 degrees.
	Tilted, single-axis tracker[B]	Concrete piers and concrete ballasts. Vendors who are able to supply utility scale projects are very limited.	5 to 15 degrees most common.
	Azimuth tracker[B]	Single-pole (metal column and concrete pier) or continuous (stem wall) foundation.	15 to 25 degrees. 30 degrees and higher also available.
	Dual axis tracker[B, C]	Single-pole (metal or concrete column and concrete pier).	0 degrees to 90 on the vertical position and +/–120 degrees on the East to West movement.

Notes
A Some manufacturers also offer seasonally adjustable structures, in which the tilt of the structure can be changed in discrete steps to better match the elevation angle of the sun. The steps are typically four in the year. No utility scale project has been built with these structures.
B Typically equipped with backtracking features.
C It is worth noting that CPV systems make use of dual axis trackers due to their requirement of tracking the daily trajectory of the sun within less than 2 degrees of accuracy. Depending on the CPV system, the system may require less than 1 degree of accuracy.

performance of the system. The designer will have to reconcile the optimal tilt and orientation derived from the results of the analytical model with the actual site conditions and mounting structure features in order to achieve an optimal balance between performance, construction practices and system cost.

The mounting structures for large-scale commercial applications can be classified into five major categories as indicated in Table 9.8. Carports are included in this table because of the potential to implement large systems when considering the vast amount of parking area available in North America. Even though carports are ground mounted systems, their performance and permitting characteristics qualify them as

Figure 9.8 Carport in a parking lot (Courtesy: Baja Construction Co. Inc).

more similar to special type rooftop systems given the interconnection requirements and the size limitations imposed by the parking lot.

With the exception of the trackers, all other structures in Table 9.8 are static throughout the life of the project once they have been installed. The exposure of the modules to sunlight is solely determined by the angle of the sun relative to the tilt and orientation of the mounting structure. The only purpose of the mounting structure is to hold the modules in the same position for the life of the project. A typical carport and ground mounted structures are illustrated in Figure 9.8 and Figure 9.9.

Photovoltaic trackers have the purpose of increasing the exposure of the modules to sunlight by following the daily movement of the sun in the sky (refer to Figure 9.2). Tracking technology enhances the output of a given PV solar field regardless of the kind of PV module technology it is paired with. If the cost of implementing a type of tracking system outweighs the production benefit of additional power generation, the developer will select the mounting option that provides the best financial project investment return. As indicated in Table 9.8, there are four main types of trackers:

1) Horizontal, single-axis tracker (HSAT)
2) Tilted, single-axis tracker
3) Azimuth tracker
4) Dual axis tracker

Figure 9.9 Ground mounted, fixed tilt structure (Courtesy: Schletter Inc.).

All trackers require a motor or a hydraulic system to enable the motion of the system, with few exceptions. Most commercial systems make use of electrical motors connected to auxiliary power lines. These auxiliary lines take power from the grid, not from the photovoltaic system.

The HSATs are the most widely used trackers in the North American market. Increased energy yields over fixed mounted systems range from 15 to 30 percent depending on the solar resource available at the location of the project. In these trackers, the modules are mounted on a horizontal structure parallel to the ground. This structure is aligned North – South in the field and it rotates from the East in the morning to the West, following the trajectory of the sun throughout the day. This rotational movement creates a better incident angle (or line-of-sight) between the modules and the sun. The maximum tilt typical of these structures is 50 degrees to the East and 50 degrees to the West. An HSAT is illustrated in Figure 9.10. The photograph was taken in the afternoon, when the sun had past the North-South meridian while traveling to the West. As the sun moves west, its elevation decreases until it reaches the sunset. For this reason, the modules are tilted to the West, and the higher the tilt, the lower the elevation of the sun. At noon, when the sun is due South and at its highest elevation, the modules are horizontal or with a zero tilt[37]. The central line is the drive mechanism (running East-West).

These trackers are a very good match for locations between the Tropic of Cancer and the Tropic of Capricorn, where the elevation of the sun is in fact perpendicular to the horizontal plane at some point in the year[38].

The HSATs are often equipped with a function called backtracking. The objective of this feature is to reduce the self-shading in the early hours of the morning or

37 Refer to section 9.2 regarding sun paths in the Southern Hemisphere and between the Tropics of Cancer and Capricorn.
38 The Tropical latitudes are approximately 23.45 degrees North and South from the Equator. Within these latitudes, the sun is perpendicular to a horizontal plane once or twice per year.

Figure 9.10 Horizontal SAT (Courtesy of PVHardware – Axone System).

late hours of the evening. During sunrise and sunset hours the elevation of the sun is low, and therefore the shadows of objects are long. If the modules were tilted to match the elevation of the sun, the length of the shadow would reach the row of modules immediately behind, as illustrated in Figure 9.11. With the backtracking function enabled, the tracker does not fully tilt to the East at sunrise, or to the West at sunset, but rather stays closer to a horizontal position in order to avoid self-shading. After sunrise, the tracker starts to rotate from a nearly horizontal position towards the East, effectively tracking the sun's trajectory in the opposite direction, or backtracking. Backtracking stops once the tracker has fully rotated to the East or the tilt has reached the best position relative to the angle of the sun. At that point, the tracker starts following the sun's path from East to West. In the late afternoon, once self-shading starts to be significant, the backtracking function is activated. The tracker then rotates from West to East, opposite to the sun's trajectory, and at sunset, the modules are nearly horizontal. Typically, backtracking is active in the morning or in the evening for one to two hours, depending on the site's latitude and the time of the year.

Tilted, single-axis trackers share the same operational principles as the HSATs, but the structures are tilted, which is particularly useful in latitudes further north or south from the Tropic latitudes. In these locations, the maximum elevation of the sun is never perpendicular to the ground and not more than 43 degrees above the horizon at Solstice (winter in the northern hemisphere and summer in the southern hemisphere). Therefore, the tilt of the trackers improves the angle of incidence (or line-of-sight) between the modules and the sun. These structures have not been used in North America as extensively as the HSATs, mostly because of the cost differential. There are a couple of large scale projects that have been implemented with these trackers.

Azimuth trackers consist of a tilted plane that rotates East to West, following the sun's trajectory throughout the day. The tilt of the plane is permanent and the axis

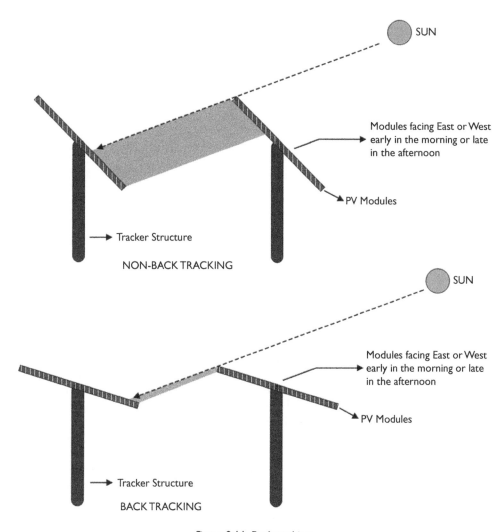

SUN

Modules facing East or West
early in the morning or late
in the afternoon

PV Modules

Tracker Structure

NON-BACK TRACKING

SUN

Modules facing East or West
early in the morning or late
in the afternoon

PV Modules

Tracker Structure

BACK TRACKING

Figure 9.11 Backtracking.

of motion is the vertical axis of the system. Therefore, all the modules on the tilted plane always have the same angle to the sun due to the rotational axis. In comparison, the modules in the tilted SAT will have slightly different angles to the sun, except at solar noon. This effect will slightly reduce energy production in the tilted SAT due to a minimal mismatch in power amongst the different modules connected to the same string. The azimuth trackers are also often equipped with backtracking. To date, these structures have not been used in large scale projects in North America. Azimuth trackers can also have a pole mounted structure, which is the same as that of the dual-axis trackers but with no elevation motion.

Dual-axis trackers, such as the one shown in Figure 9.12, consist of an array of modules supported by a single pole. Increased energy yields over fixed mounted

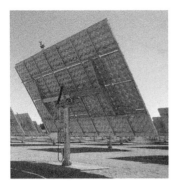

Figure 9.12 Dual axis tracker and CPV System.

systems are said to be in the range of 25–45 percent depending on the solar resource available at the location of the project. The mechanical assembly of this tracker is integrated by two sets of motors, one that generates an East to West rotation and another that enables an up and down motion. This double motion of the tracker (dual-axis) enables the system to accurately follow the sun's path throughout the year. In this manner, the modules have a direct line-of-sight to the sun at all times, or in other words, the modules have the maximum amount of "fuel" available at any time during the day and throughout the year. The degree of accuracy depends on the control mechanism used by the tracker manufacturer, but only concentrating photovoltaic systems (CPV) require a high degree of accuracy to function properly. In these cases, the motion of the system is almost continuous. Azimuth trackers can have this same structure, with the array at a fixed tilt and only one motor installed in order to enable the East to West rotation.

To date, dual-axis trackers have seldom been used in North America for large scale projects using flat-plate modules. However, there are a significant number of projects in Spain that were implemented with these trackers. The market/policy environment in Spain benefited the installation of these trackers due to the feed-in-tariff structure, in which energy contracts were paid out based on energy output. Therefore, project developers would design a PV system maximizing energy output, which resulted in the use of dual axis systems. Another factor that played into decisions of utilizing dual axis tracker types was that solar PV modules prior to 2009 were significantly more expensive than they are in 2012, so developers had to maximize the production capability of each module to the best of its potential. Considering that CPV systems require the use of a dual-axis tracker, there a number of important projects in North America using dual-axis trackers that have already been completed or are in development.

In general, the use of trackers in North America for large scale projects has been limited as of the end of 2011. Most systems have been installed with fixed tilt mounting structures. The large majority of photovoltaic projects developed with trackers have used horizontal, single-axis trackers with a notable exception: One of the first utility scale projects in the United States was implemented with tilted, single-axis trackers[39]. There are very few large commercial projects that have been implemented with dual-axis trackers in North America. From the point of view of reliability, durability and engineering, the most significant features of a mounting structure are, in general:

a) Structural integrity: Considering that the mounting structure is holding all the modules of the system, the design and construction of this structure, in conjunction with the foundations, must be able to support the full dead weight of the modules. The typical weight of flat plate modules (any flat-plate technology) is between 11 to 16 kg per square meter. More importantly however, the complete mounting system should be able to support environmental loads with modules installed. Wind loads are the most relevant specifications of the mounting structure, with standard wind exposure of 90 miles per hour (145 km/hr). Depending on the site location, higher wind ratings can be required (e.g. hurricane zone) as well as seismic loads and snow loads, with additional requirements for the foundations of ground mounted systems such as the lateral pressure of soil, ground water and flood, corrosive soils and other specific site conditions.

The structural integrity of the mounting system can be evaluated with analytical calculations and with the aid of computer tools such as finite element analysis (FEA). Structural testing, such as wind tunnel testing or highly accelerated life testing (HALT), on components and the system, as well as field data, will complement the analytical findings and provide further insight into the structural integrity of the system. Although these tests increase confidence of the analytical results, they are not always performed due to the associated costs or lack of long term operational history.

The designer should be aware of extreme environmental loads that are likely to occur at the project's location within the life time of the project (20 to 25 years) and should request structural integrity analysis fitted to those conditions.

b) Assembly: The reliability of the system assembly will be determined by the structural integrity analysis. However, from an engineering perspective, the designer should understand the activities and resources (personnel and equipment) involved in the assembly and construction of the system on-site. The designer should assess the degree of complexity involved in the assembly and installation processes and the impact this can have on the construction schedule and installation costs.

c) Quality of materials: The long term durability of the system is dependent on the structural integrity and the quality of the materials used to assemble the system. The designer should be aware of the environmental factors that are likely to accelerate the degradation of the system's materials over the life time of the project such as corrosive soils, saline fog, high humidity, high temperature, sand and others. The designer should provide this information to the manufacturer and ask about the options to mitigate these effects. According to the specific site conditions, the materials and parts used in the system should meet recognized quality and performance (if applicable) standards. The list of standards and a copy of qualification testing and certifications should be provided by the manufacturer. Metal standards widely recognized by the industry in North America have been

39 The 14 MW system installed at the Nellis Air Force in 2007.

developed by ASTM international, formerly known as the American Society for Testing and Materials. The European Standards (EN) are also widely recognized and accepted. Standards developed by other organizations may be equivalent to ASTM and EN standards. The designer should judge the applicability of the standards provided by the manufacturer in relation to the specific parts and uses of the system, the testing laboratories used for certification and the recognition and acceptance of those standards by the permitting authorities.

d) Reliability program: In addition to the structural integrity analysis, the designer should ask about any further analysis or testing of individual parts or subsystems performed by the manufacturer in order to detect and evaluate failure mechanisms, mean time between failures, expected wear and tear over the useful life of the system, performance under high stress conditions, long term reliability and ultimately establish a lifetime expectation and the basis for that expectation.

9.4   PHOTOVOLTAIC SYSTEM DESIGN

This section describes the integration of the major components reviewed so far in this chapter into a photovoltaic system.

A grid connected and large-scale photovoltaic system can be seen as composed of a DC field and an AC field, each one relatively independent of the other. The DC field is comprised of electrical and mechanical components. The main electrical components include the modules, inverters, wiring, combiner boxes, grounding and lighting protection, all associated breakers, disconnects, fuses, and other electrical safety devices. The mounting structures and enclosures are the main mechanical components. The AC field is comprised of a step-up transformer, wiring and protection devices. Grounding and lighting protection are typically integrated within the DC field.

A photovoltaic system is a relatively simple electrical circuit, although the large number of components and variables involved in its design can present interesting engineering challenges. Figure 9.13 is a block diagram of a typical utility scale photovoltaic system integrated with a central inverter system.

In general, the detailed design of a photovoltaic system involves the next items: operating voltage, temperature, string sizing, combiner box, DC wiring, DC to AC conversion, AC field, and system modeling.

9.4.1   Operating voltage

The maximum operating voltage of a photovoltaic system is determined by local, state and/or national regulatory agencies. In the United States and Canada, the National Electrical Code's (NEC) and the Canadian Electrical Code's (CEC) definitions have been used as regulatory bodies for photovoltaic systems, in conjunction with specific standards and certifications that equipment and materials used in photovoltaic systems must meet. In these countries, PV systems are for the most part limited to a maximum operating voltage of 600 V_{DC} although there is a trend to allow higher voltages for utility scale projects. The same amount of power transmitted at a higher operating voltage results in a lower current density carried by the conductors. The transmission of electrical power obeys $P_{ELECTRICAL} = I \times V$ thus, increasing the voltage

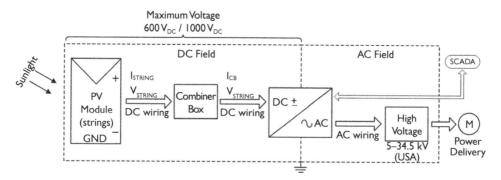

Figure 9.13 PV system (Block Diagram).

from 600 to 1000 (or 1.67 times) will deliver the same power at 0.6 the value of the current. Because the power lost as heat in the conductor is proportional to the square of the current ($P_{HEAT}=I^2 \times R$), but does not depend in any major way on the voltage, increasing the voltage in a power system reduces the conductor losses per unit of electrical power by a factor of I^2. Since the resistance depends on the material, diameter and length of the conductor, increasing the voltage on a conductor of the same material will allow either a) an increase in the current, b) a reduction in the diameter, or c) an increase in the length, while delivering the same amount of power. This property is mostly used to specify conductors with a smaller diameter thus, saving in cable costs. Aside from the reduction in resistive losses, there are also a few other design, construction, and cost advantages to using a higher operating voltage.

The NEC is published by the National Fire Protection Association (NFPA). The NEC was approved as an American national standard by the American National Standards Institute (ANSI) and it is formally identified as ANSI/NFPA 70. The NEC is updated and published every three years with the 2011 NEC being the current edition. The Canadian Electrical Code pertaining to the installation and maintenance of electrical equipment in Canada is published by the Canadian Standards Association (CSA). The document is formally identified as CSA C22.1 and is also updated every three years. The technical requirements of the CEC are very similar to those of the NEC but specific differences do exist. There is an ongoing effort to correlate the technical requirements between the two codes. In a larger context, several electrical equipment standards have been harmonized between the USA and Mexico. The Council for the Harmonization of Electromechanical Standards of the Nations of the Americas (CANENA) is working to harmonize electrical codes in the western hemisphere. For the purposes of this discussion, it is assumed hereinafter that the sections of the NEC and the CEC applicable to photovoltaic systems are compatible.

The NEC is a compilation of recommendations that are considered minimum provisions to design and install a safe electrical system. The purpose of the NEC is the practical safeguarding of persons and property from hazards arising from the use of electrical installations. NEC recommendations are typically mandated by state or local law and most states adopt the most recent edition within a couple of years of its publication. However, the local authority having jurisdiction (AHJ) is the entity that ultimately inspects and enforces the compliance of each particular installation to the NEC. Consequently, the AHJ is responsible for making interpretations and enforcing the requirements of the NEC. On rare occasion, the AHJ will grant exceptions to the NEC or special permissions.

The requirements of the NEC are meant to be applicable to all public and private premises that can be accessible to the general public or to people not trained in electrical safety procedures (e.g. houses, buildings, parking lots, mobile homes, warehouses, machine shops, garages, etc).

The NEC includes provisions to regulate the installations of photovoltaic systems in Article 690 of the code. Amongst these provisions, the code defines the maximum DC voltage of a photovoltaic system to be 600 V_{DC}. Even though Article 690 includes the possibility of installations with voltages over 600 volts (covered in Article 490, which is not specific to photovoltaic systems), the 600 V_{DC} limit has been typically considered as the maximum allowable voltage for all photovoltaic installations where the NEC applies. This limit has not been challenged in a significant manner in the several past editions of the NEC, when the photovoltaic industry was still confined to small systems. However, between the 2005 and 2008 editions of the NEC, the size of photovoltaic systems grew dramatically and several large scale systems began to be installed. In the era of the current 2011 edition of the NEC, multiple large scale projects have been completed in North America and a significant number of projects are under construction or development. The sizes of these projects, as well as the experience gained from the photovoltaic industry in Europe (where 1000 V_{DC} systems are the norm), have been challenging the traditional 600 V_{DC} limit recommended by the NEC and enforced by the local AHJ. However, most AHJs prefer to follow the established rules of the code as well as require UL or NRTL certified equipment and materials[40].

Notwithstanding, the NEC is not meant to be used as a regulatory code in several instances. For the specific purposes of photovoltaic system design and installation, there are two relevant instances where the NEC is not necessarily enforced:

a) In installations under the exclusive control of an electric utility, where such installations are on property owned or leased by the electric utility for the purpose of electrical generation.

b) The authority having jurisdiction to enforce the NEC can also grant exceptions to systems connecting to service conductors, when these systems are outside a building or structure, or terminate inside the nearest point of entrance of the service conductors.

These exceptions have been used when applicable, to gain approval from the authorities for systems with an operating voltage greater than 600 V_{DC}. According to the first item above, these systems have to be installed within premises with restricted access and treated similarly to traditional electrical generation facilities. To date, these systems have been installed in open spaces and a fence, or other physical barrier to restrict access, typically surrounds the whole project site. Because of this, these systems are known as "behind the fence" photovoltaic systems. A "behind the fence"

---

40 Underwriters Laboratories (UL) and Nationally Recognized Testing Laboratory (NTRL).

system does not strictly imply that the system has an operating voltage greater than 600 V_{DC}[41], it only means that the system has restricted access from the electrical safety point of view and that the meter is not installed on the utility side of the point of interconnection. The authorities granting the permits for construction and operation of the facility ultimately decide if a behind the fence system should or should not comply with the 600 V_{DC} limit and other requirements (such as third-party validation of components and materials, i.e. listing or certification). This has been the case for many large scale ground-mounted systems in Canada and the United States, where behind the fence systems have had to be designed and built as 600 V_{DC} systems, either following compliance with NEC limits imposed by the authorities or because equipment and materials meeting the authority's requirements (such as UL or CSA listing) were not rated for applications with an operating voltage greater than 600 V_{DC}.

In summary, the design and installation of photovoltaic systems in the US and Canada has been largely defined by the requirements specified in the NEC and CEC and enforced by the AHJ. Currently, the NEC and CEC requirements may not be enforced only for behind the fence applications. The designer must be aware of potential or actual restrictions regarding the maximum operation voltage of the system as this will define important characteristics of the system.

9.4.2   Temperature

The ambient temperature expected on site is an important parameter to consider for PV system design. This is because the voltage of the PV modules is an inverse function of the cell temperature and the maximum voltage of the system has well defined limits as seen in item 1.5.1. Figure 9.14, left side, shows the change in voltage as a function of the cell temperature at a fixed irradiance value. The irradiance graph on the left side in Figure 9.14 is 1000 W/m² and the module is a typical poly-crystalline silicon version rated at 285 W with an open circuit voltage (V_{OC}) of 44.8 Volts at STC. This graph shows that an increase in cell temperature (curves to the left) corresponds to a reduction in V_{OC} and a reduction in power output. The inverse occurs with a decrease in cell temperature, which is why if a PV module would be cooled, the efficiency would increase.

Although the change in voltage is not a linear process, in practice it can be approximated within a reasonable range of temperatures by the linear function as follows:

$$V_{OC}(T) = V_{OC} + \alpha V_{OC} \times (T - 25) \tag{1}$$

Where $V_{OC}(T)$ is the open circuit voltage of the module at a cell temperature of T and V_{OC} is the open circuit voltage at STC. The temperature coefficient factor for open circuit voltage, α, is often expressed as a percentage rate per degree Celsius. A typical value for crystalline silicon modules is −0.32%/°C.

The graph on the right of Figure 9.14 shows the change in voltage as a function of cell temperature at a lower irradiance of 200 W/m². Equation (1) can also be used to calculate $V_{OC}(T)$ if the V_{OC} at the new irradiance conditions and cell temperature of 25°C is known.

---

41 In fact, most utility scale systems installed in Canada (including the 80 MW project at Sarnia) are behind a fence and considered electric generation facilities, but the maximum operating voltage of those systems is 600 V_{DC}.

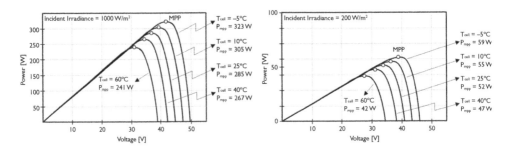

Figure 9.14 Module voltage as a function of temperature.

There are two important characteristics to note in the graphs of Figure 9.14:

1) The V_{OC} is also dependent on the irradiance level. A higher irradiance increases the voltage and the inverse occurs with a lower irradiance. This relationship can also be approximated as a linear function. For instance, the V_{OC} of the module at 1000 W/m² and 25°C cell temperature (purple line, Fig. a) is 44.8 V_{DC}, while the V_{OC} at 250 W/m² and 25°C cell temperature (purple line, Fig. b) is less than 41 V_{DC}.

2) The cell temperature-as the ambient temperature- is a direct function of the irradiance level on the cell. The graphs in Figure 9.14 do not show this relationship. These graphs only show the V_{OC} as a function of the cell temperature at a fixed irradiance level without considering the thermal gain of the cell due to the irradiance level. When exposed to sunlight, the cell temperature is above ambient temperature due to the sunlight absorption properties of the cell, the insulation provided by the protective layers and the conversion inefficiencies of the cell.

In summary, the V_{OC} depends on the cell temperature, which in turns depends primarily on the irradiance level incident of the cell. In regions with ambient temperatures below freezing, the cells can take one or two hours before gaining enough heat to set them above ambient temperature. High speed winds and low irradiance levels will further delay the heat gain of the cell, which will then maintain a temperature similar to ambient.

The industry standard is to use the calculated V_{OC} (T) at low temperature as the maximum operating voltage defined for the system, either 600 V_{DC} or 1000 V_{DC}. The value of this V_{OC} is calculated using equation (1). The value of T in equation (1) refers to the cell temperature, which, as said before, depends primarily on incident irradiance, and, to a lesser degree, heat transfer mechanisms. Rather than calculating the cell temperature, equation (1) is used to assume that the cell temperature is the same as ambient temperature. This approximation simplifies calculations and is reasonable for the initial exposure of the cell to sunlight. One drawback of this approach is that the reference V_{OC} in equation (1) is based on an irradiance level of 1000 W/m², which won't be the case when initial exposure of the cell to sunlight is at the beginning of the day. This assumption inherently introduces a safety margin by calculating a higher V_{OC} (T) than actual, which in some instances can be challenging due to the MPPT window of the tracker.

Table 9.9 Correction factors for VOC (NEC Table 690.7).

Minimum temperature (°C)	Correction factor
0 to 4	1.10
−1 to −5	1.12
−6 to −10	1.14
−11 to −15	1.16
−16 to −20	1.18
−21 to −25	1.20

In addition, there is no standard to determine the ambient temperature, T, to use in equation (1). It is common practice to observe historical meteorological data and based on this information, select one of the lowest values in the trend. It is important to note that this temperature value should be selected at sunrise, where the lowest temperature during daylight hours is likely to happen and when the cell and ambient temperature are the closest. The selection of this temperature tends to be made conservatively, which further accentuates the already biased effect of using the reference V_{OC} at 1000 W/m².

With this method, the calculated value of the maximum expected V_{OC} (T_{min}) at the lowest historical temperature provides enough margin to ensure that the PV system voltage will not exceed the maximum operating voltage at any time.

For instance, the 285 W module used in the graphs of Figure 9.14 has a temperature coefficient for V_{OC} of −0.335%/°C and a V_{OC} of 44.8 V at STC. Assuming a lowest temperature of $T_{min} = -5°C$, the V_{OC} (T_{min}) is 49.3 V (left graph of Figure 9.14). However, the V_{OC} at 200 W/m² is approximately 40.8 V thus, the V_{OC} (T_{min}) under these conditions is 45 V (right graph of Figure 9.14).

The temperature coefficient factor varies slightly within crystalline silicon and thin film technologies. The same equation is used for both.

The NEC follows an even more conservative although simpler approach. Table 690.7 of the code has correction factors for 10 different temperatures intervals. Table 9.9 presents the correction factors for the first six intervals.

Applying this table to the previous example, the calculated $V_{OC}(T_{min})$ would be 51.1 V. Following NEC rules, the appropriate correction factor to use is 1.14. This table is intended for crystalline silicon and thin film technologies.

9.4.3   String sizing

Modules are connected in series to form a string. Similar to a battery connection, the voltage of the string is the sum of the voltage of each module. The current of the string is the same as the current of one single module. The modules are connected using the cables and connectors (one positive and one negative) included with the module. The number of modules per string is defined primarily by the maximum operating voltage of the system, 600 V_{DC} or 1000 V_{DC} for utility scale projects. Since the open circuit voltage is the maximum voltage that a PV system can achieve, the V_{OC} (T_{min}) is used as the limit for the maximum operating voltage defined for the system. For crystal-

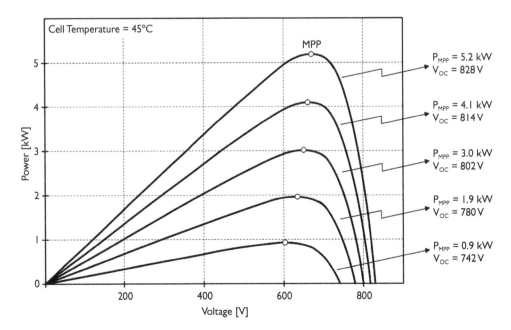

Figure 9.15 Maximum power point as function of system voltage.

line silicon modules, the typical string size varies between 12 to 24 modules while for thin film technologies, the typical string size varies between 8 and 18 modules. There is no typical module size for CPV systems, but they are almost always able to meet the maximum operating voltage requirement. The typical string current is less than 10 A_{DC} across all technologies.

The size of the string has a direct effect on the performance of the system due to the MPPT window of the inverter. This characteristic refers to the system voltage window where the inverter is active and converting DC power to AC power by setting the system at the maximum power point. Figure 9.15 shows the maximum power point as a function of the system voltage at different irradiance levels. The cell temperature is fixed at 45°C for all curves.

By design, the inverter requires a minimum voltage input to turn-on and an upper limit beyond which the MPP tracking algorithm stops. For 600 V_{DC} inverters, this window is approximately 300 to 600 V and 450 to 850 V for 1000 V_{DC} inverters. The precise window is design specific and varies amongst manufacturers. When the system voltage is below the minimum MPP voltage, the inverter will remain in stand-by mode, not generating power. If the system is above the MPP voltage, the system will shut down for self-protection or will set an operating point at a fixed voltage and will resume MPP tracking only at lower voltage levels[42].

---

42 This situation would be rare as it would require low temperatures and high irradiance levels. In most cases where the string voltage exceeds the maximum MPP window, the string has not been properly sized.

Table 9.10 String sizing.

| Case | $V_{OC}(T_{min})$ | String size (# modules) | | String voltage ($1000\ V_{DC}$) | |
		$600\ V_{DC}$	$11000\ V_{DC}$	$V_{MPP}/V_{OC}^{\ A}$	$VV_{MPP}/V_{OC}^{\ B}$
1) 1000 W/m²	49.3	12	220	750/910	6664/800
2) 200 W/m²	44.9	13	222	825/1001	7730/880
3) NEC (1.14)	51.1	11	119	712/865	6630/760

Notes
A V_{MPP} and V_{OC} at STC (datasheet, typical, $V_{MPP} = 37.5\,V$; $V_{OC} = 45.5\,V$).
B V_{MPP} and V_{OC} at 75 W/m² and 15°C cell temperature (From software model, $V_{MPP} = 33.2\,V$; $V_{OC} = 40\,V_{VOC}$).

Therefore, the optimal size of the string should cover the MPP range without exceeding the maximum voltage ratings of the inverter and project regulations.

A conservative approach can lead to a short string that can frequently set the system voltage below the lower limit of the MPPT window. This translates into energy losses by delaying the DC to AC conversion of the inverter until higher irradiance levels increase the voltage of the string. Meanwhile, the power generated by the strings is not used. A string with too many modules will also translate into energy losses by setting the operating point of the string at a lower point than maximum. This later case is less likely to occur, as the maximum string size defined in terms of the V_{OC} (T_{min}) typically covers the upper limit of the MPPT as well.

Table 9.10 provides an example of string sizing under three different criteria. This Table is based on a 285 W module (44.8 V_{OC} at STC) for a $T_{min} = -5°C$.

The typical method used for design is either case 1 (STC) or case 3 (NEC). Case 2 may present permitting challenges due to V_{OC} exceeding the maximum operating voltage. As discussed in the previous section, an analysis of the irradiance and weather patterns specific to the location will provide a clearer correlation on realistic irradiance levels and lowest cell temperatures expected on site.

Once the string size has been defined, the operating voltage of the system is defined as the operating voltage of the string.

9.4.4 Combiner box

This component has the function of combining a set of strings by connecting them in a parallel form. This reduces the number of wires running from each string to the inverter. The voltage of the combiner box will be the same as the string voltage, and the output current of the combiner box will be the same as the sum of the current of each string. The typical current capacity of the combiner boxes is 200 A_{DC} although capacities up to 400 A_{DC} are not uncommon. For crystalline silicon modules, a 200 A_{DC} combiner box is used to connect approximately 16 strings, while the same box would connect approximately 60 thin film module strings. The combined power of the strings is sent to the inverter with a cable appropriately rated for the current output. The combiner boxes integrate some protection devices such as fuses at each input, and breakers and DC disconnects in some cases. Combiner boxes can also include instruments to monitor current and voltage at each input, and report to a SCADA system.

9.4.5   DC wiring

This includes all the connectors, terminals and cables necessary to build strings, connect them to the combiner box, and connect the output of each combiner box to the inverter. All modules include a set of cables and connectors as part of their design and specifications. The typical cable size is 4 to 6 square millimeters, or 10 to 12 AWG terminated in a copper, tin-plated contactor crimped to the cable, and covered by the connector's structure. Most module manufacturers use a type of connector originally developed by Multi-Contact AG referred to as MC4. Other companies use an equivalent or similar type of connector. These connectors increase human safety and the reliability of the product by providing a snap and lock connection with a non-exposed contactor, water tight seal, unique polarity ("female" and "male" types) as well as UV resistant and long lasting plastic material. These module integrated cables are used to build strings. Once a string is completed, a cable dimensioned to minimize electrical losses due to conductor length attenuation covers the distance from the end of the string to the location of the combiner box. The dimensions are usually the same or slightly higher (in diameter) than the module cable, between 8 to 12 AWG. Due to the low current of each string in the case of thin film modules, it is common practice to use a custom made cable harness that combines several strings before connecting it to the combiner box. The specifications of the module cable, string cable, cable harness and any other type of cable or connectors have to meet outdoor conditions (UV, moisture and heat resistant material and construction) in order to ensure the high reliability of the system and last for the expected 20 or more years of operation in the field. In some installations, particularly rooftop systems, metal conduit is used to further protect these cables from the environment and for fire safety.

The power from the combiner box is then routed to the central inverter. The distance from the combiner box to the central inverter can be significant in terms of voltage drop and thus, the size of the cable has to be selected accordingly to minimize power losses. The maximum voltage drop allowed due to cable length is generally less than 2 percent at maximum load. The current carrying capacity (or ampacity) of these cables needs to match the ratings of the combiner box output. These cables must also meet the DC voltage rating of the string. The cables are often buried in the ground from the combiner box to the central inverter to avoid the use of a duct bank and reduce costs. The specifications of these cables need to address installation methods such as direct burial, outdoor or others.

Although the use of a central inverter is most common, there are a few other system configurations such as the use of distributed inverters or DC to DC converter stages before connecting to the inverter. In either case, the specifications of the DC wire need to address the voltage drop, ampacity and operating voltage requirements of the system, as well as the installation methods. In addition, minimal quality and safety standards, as well as permitting requirements, should be kept in mind when selecting characteristics and suppliers.

9.4.6   DC to AC conversion

The power conversion is performed by the inverter, which is also enables other important control operations as described in Section 9.4.2.

The power output is directly proportional to the power generated by the DC field minus the energy losses throughout the system stages. The power generated by the DC field is a function of the irradiance and the number of strings connected to the inverter. The number of strings is generally defined by one or more of the following aspects: space constraints, power output profile, and optimization.

a)   Space constraints: This situation is more likely to occur in rooftop systems where the number of modules depends on the space available. In this situation, the inverter has to be chosen based on the capacity set by the number of modules.

b)   Power output profile (short and long term): The DC power profile follows the daily irradiance profile as modified by the orientation and type of mounting structure (fixed tilt or tracker). This profile has the general shape of a bell curve as illustrated in Figure 9.16 (green and blue line). However, the DC capacity connected to the inverter can change this profile by widening and flattening the top of the curve (purple line).

If the DC capacity is the same as the inverter capacity, it is highly unlikely that the inverter will ever reach maximum capacity. This is because, as discussed in section 1.4.1, the conditions to define capacity in W_p are not representative of the field conditions. Instead, the PVUSA Test Conditions (PTC) or Nominal Operating Cell Temperature (NOCT) are better suited for this. Using PTC, the field capacity of the DC system at 1000 W/m² is approximately 90 percent of the W_p capacity, depending on the specific technology and module. This means that

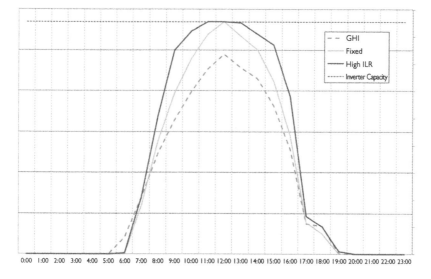

Figure 9.16 Energy output profiles.

to reach the nameplate capacity of the inverter, the DC field has to have a W_p capacity at least 10 percent larger. This is commonly expressed as an "Inverter Loading Ratio" (ILR) of 1.1. This term refers to the DC field capacity (W_p) to the inverter nameplate AC capacity (W_{AC}) ratio and it is a widely used definition. Even with an ILR of 1.1, the system may reach the inverter capacity only at irradiance levels of 1000 W/m² or higher, which are not common. For the 1.1 ILR fixed tilt system in Figure 9.16 (light line) this only occurred once in that particular day. As this graph shows, the full capacity of the inverter (dotted line) is under-utilized. In order to take full advantage of the maximum capacity of the inverter, ILRs of 1.15 to 1.35 are typically used. The ratio selected depends on the irradiance conditions on site and the specific module technology. A system with a higher ILR (dark line) shows that the inverter was at full capacity for a few hours at the peak of the day. In this case however, the inverter is "clipping" the top of the curve so as to not exceed its power rating. The extra energy above the maximum inverter capacity is lost. The extra energy generated (in comparison to the 1.1 ILR system) by widening the curve with increased DC capacity may compensate for the energy lost. Given the seasonal variations of solar resources, the inverter will still operate below its maximum capacity for a significant part of the year. The limit for ILR is defined by the maximum input current of the inverter at a given voltage. This limit has to be verified with the inverter manufacturer. Most utility scale inverters can remove the clipping feature and allow for the extra generation of energy or, if reactive power supply is enabled, this extra energy can be used to provide reactive power without losing real power.

In the long term, all photovoltaic technologies degrade and the power output decreases year after year. For crystalline silicon, this loss in power is equivalent to approximately 0.7 percent per year. Considering this factor, the ILR can also be increased to maximize the inverter capacity even after several years of operation.

c) Optimization: The previous item discussed how increasing the number of strings can flatten the output profile. This relationship is impacted by other variables such as orientation, tilt, irradiance, operating temperature of the DC field and the inverter (which also relates to ILR), clipping, maximum input current, losses from the field to the inverter, and reactive power requirements. In order to gain a system wide perspective that takes into account the above-mentioned variables and others, it is necessary to build a model of the system. This model can be built with the use of in-house developed software tools and/or with readily available programs for photovoltaic system analysis[43]. With the help of these software tools, a well developed optimization goal, system constraints, and methodology, it is possible to analyze the multi-variable effects on the behavior of the system. This will allow for a variable control and sensitive analysis in order to optimize the output profile. The optimization will lead to an optimal definition on the number of strings (ILR), orientation, tilt, and other specific variables.

During this optimization process, it is also important to include the capital costs implicit in each iteration as well as the energy production and revenue generated. The most optimal design from the engineering perspective may not necessarily translate into the most optimal design from the financial perspective.

---

43 For example, PVsyst, PVSol, SAM, RETScreen, Insel, TRNSYS, PV F-Chart, Solarpro, MauiSolar, etc.

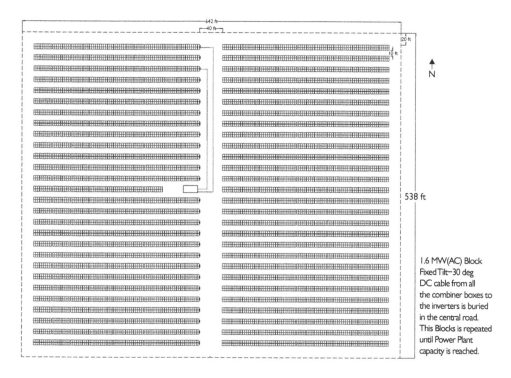

Figure 9.17 Building block.

The most prevalent arrangement of the inverters in the DC field consists of blocks between 1 MW$_{AC}$ to 1.8 MW$_{AC}$. Each block is integrated by a set of inverters, typically one or two, physically located next to each other and generally in the middle of the DC field. Each block is a self-contained photovoltaic system that connects to the other blocks in the power plant on the AC side. The concept of building blocks consists of developing one basic design and aggregating it block after block until the total power plant capacity is achieved, without modifying the electrical design or construction methodology. This approach allows for the development of a set of repetitive procedures and operations in design and construction. In this manner, the design time can be reduced and the use of materials and labor can be optimized during construction. The effect sought is a reduction in equipment and construction, as well as higher productivity and accelerated construction time. Figure 9.17 illustrates the building block approach.

9.4.7   AC field

The output of the inverter is a grid quality signal, and conventional AC practices are used to connect the inverter to the grid. The typical arrangement of the AC field consists of two inverters per block connected parallel to a distribution transformer. This transformer steps-up the output voltage of the inverter (typically between 250 V$_{AC}$ to 380 V$_{AC}$ for 1000 V$_{DC}$ inverters) to a standard distribution voltage in order to reduce wiring losses. In the United States these voltages are 4.16 kV, 12.47 kV, 13.2 kV, 14.4 kV, 24.9 kV,

19.9 kV, and 34.5 kV. Some of these voltages are more prevalent than others. The standard approach is to route the AC output of the transformer via direct buried cables to one or more central switchgears, depending on the design of the AC network topology. No specific topology is preferred over others. The cables must be properly sized and installed. The AC field must also include a grounding and lighting protection circuit.

Another relevant aspect to note is the monitoring and control system. With the expansion of photovoltaic systems in the distribution and transmission system, as well as advances in smart grid technologies, there is a trend towards more structured monitoring and remote control of these systems. Most large scale inverters are already capable of sending information and receiving control commands through an external SCADA system. Currently, the control requirements and control capabilities are rather basic. It is expected that in the near future, control requirements will increase in complexity to address the characteristics described in section 9.4.2. Therefore, it is important for the designer to discuss with the inverter manufacturer the control and SCADA integration capabilities of the inverter.

9.4.8   System modeling

The performance of a photovoltaic system can be analyzed by a mathematical model (a set of equations and solving methods) that describes the behavior of an actual system. The importance of a PV system model and PV model analysis is two fold:

a)   Design optimization
b)   Economic variable

In general, there are two main approaches to model a photovoltaic system, the one-diode model and the parametric model, the later based on measured PV modules or system parameters. There is a significant amount of literature on PV system modeling with varying degrees of complexity and precision. The best approach to solve these models is with the use of a computer program that can be developed in-house or purchased. More complex models are a function of several variables that can be organized in the categories described in Table 9.11.

In general, Table 9.11 captures the most significant variables of a PV system model although, depending on the complexity of it, a few other variables can be added or removed.

Considering the amount of variables that affect the performance of a PV system, it can be seen that the use of an accurate software model can greatly help optimize the output of an initial design. A computer program can create and solve multiple scenarios in a short time by feeding information to an optimization algorithm. This analysis will be able to guide the designer to the optimal solution sought to address a specific output goal under specific system constraints.

PV system models are also a valuable tool in finalizing the economic analysis of the project. One immediate implication of the design optimization process is the effect it has on capital, operation, and maintenance cost. For example, the performance analysis of a fixed tilt versus a tracker system will generate a set of different energy output values that will be accompanied by a different set of project costs due to different construction, equipment, and maintenance requirements.

Table 9.11 PV system model.

Category	Variable	Description
Input	Irradiance	Irradiance is the "fuel" of the system, and, in general, the only input to the system.
Components	Module, inverter and other components (batteries, transformers, etc)	Parameters and functions representing the performance of the components.
Gain	Concentration, albedo, temperature, orientation, tilt	The orientation and tilt can increase or decrease the energy generation capacity of the system. Other parameters impact the performance of components.
Loss	Shading, soiling, resistive losses (wiring and transformers), inverter efficiency, module mismatch, module quality, thermal effects, degradation	Shading: shadows over the module created by system objects or distant objects.
		Soiling: Dust, snow, bird droppings or other substances that deposit on the surface of the module and reduce sunlight input.
		Module mismatch: Differences in performance amongst modules connected to the same string.
		Module quality: Power variability amongst modules.
		Thermal effects: High operating temperatures of the module and inverter reduce power output. Ambient temperatures are directly correlated to irradiance levels.
		Degradation: Inefficiencies accumulated in the long term due to degradation of the PV module and inverter materials.
Output	Electrical energy	The amount of energy generated by the PV system taking into account all of the above variables. Typically, models generate hourly values.

The results of the cost analysis of each design scenario, in conjunction with the output profile of each design, and the economic constraints and energy profile requirements of the project, can be used to select the most cost effective solution.

From the perspective of a project owner or investor, the most relevant variable of the PV power plant is the electrical energy that will be generated according to the specific demand profile. In utility scale projects, the goal is often to generate the maximum amount of energy on an annual basis, but sometimes meeting certain time of day schedules are more profitable than maximum annual generation designs.

Therefore, the PV system model becomes a tool to estimate the energy that will be generated by a specific design. This output is one of the main inputs used in the financial model to calculate the revenue that the project will generate. Hence, it is important to use accurate models and computer programs in order to generate trustworthy estimates. However, given the variability of solar resources, even the most accurate model will have an uncertainty band. Understanding the width of this uncertainty band is also key to developing a robust financial model.

Preface: Business section

Solar energy is a business like any other industry. While at times it is very technical and complex, the foundational aspects of business, finance, and investment are valuable to understand. While slight differences exist in the technology used or legal requirements of a region, the business topics (especially topics related to the law, risk management and finance) addressed here are relevant to almost any type of energy development project in any part of the world. As the solar energy industry is exemplary of the global economy, having equipment manufactured in one corner of the world and installed in the opposite, a global perspective is key to success. The business section and especially the chapters relating to current and future markets will provide the reader with not only predictions, but also examples of successes and failures in renewable energy markets. This information is invaluable to both entrepreneurs and policy makers alike.

The following is a brief list of this section's chapters and highlights of the contents:

Legal Considerations of Solar Development by *Jeff Atkin and William DuFour III*
Finance by *Michael Mendelsohn*
Insurance and Risk Management by *Scott Reynolds*
Mexico – An Emerging Market with Promise by *Pilar Rodríguez-Ibáñez*
The Future of the US Solar Market by *Brett Prior*

Chapter 10

Business: Legal considerations of solar development

Jeffery R. Atkin and William DuFour III

This chapter is intended to provide the reader with a general background of certain legal terms and key documents commonly used in the development of solar energy projects. This chapter is divided into the following three sections: (I) Entering Into Business Negotiations, (II) Business/Deal Structuring Decisions, and (III) Project Development Considerations. However, these sections are not necessarily sequential. In other words, some of the points made in all three sections will be relevant in most, if not all, stages of project development.

The contents of this chapter are not meant to replace the use of legal counsel; rather, we intend to help the reader identify and understand important deal points and common considerations in the documents described within this chapter. In practice, the "key terms" sections should be used as checklists (or cheat sheets) when reviewing certain documents. All to say: this chapter is not meant to be a comprehensive review of the agreements described, but is meant to provide practical considerations that the reader should have in mind when reviewing various documentation. With that said, every project developer should consider engaging experienced counsel throughout project development because expertise is required in the following areas: real estate; environmental and siting; partnership structures; Securities and Exchange Commission (SEC) controls, as they relate to holding companies; Federal Energy Regulatory Commission (FERC) regulation; state regulatory authority energy regulation; utility contracts, including power purchase agreements, operation and maintenance agreements, and interconnection agreements; construction contracts; project financing; tax issues; and evolving market trends.

10.1 ENTERING INTO BUSINESS NEGOTIATIONS

A project developer will enter into business discussions with various parties throughout the stages of project development. Two common documents that may be important to some or all of these discussions are:

1) a confidentiality agreement, in which the parties agree to keep certain information confidential, and
2) a term sheet, in which the parties outline the agreed upon deal points that will be central to their later agreement.

10.1.1   Confidentiality agreements

A Confidentiality Agreement, also referred to as a Nondisclosure Agreement or "NDA", is a contract through which the parties agree not to disclose information covered by the agreement. NDAs are commonly signed early in initial talks and/or negotiations between two parties – when they are considering doing business and need to understand the processes used in each other's business for the purpose of evaluating the potential business relationship. Thus, an NDA protects non-public business information. NDAs are useful because they help the parties define what information they consider to be "confidential." They set the expectations of the parties, and they may cover additional issues such as non-solicitation.

Key Terms in a Confidentiality Agreement:

- **Parties** – Consider whether the affiliates of one or both parties should be covered by the agreement.
- **Business Purpose** – Many NDAs limit the disclosure or exchange of confidential information for a specific business purpose (e.g., "to assess a potential panel supply agreement between the parties"). This is helpful in drafting access and use restrictions (e.g., a recipient may share confidential information among its affiliates, representatives, employees and contractors only for the stated business purpose).
- **Definition of Confidential Information** – It is important that confidential information be defined broadly enough to cover all information that may be disclosed, without being overly broad and circular. Also, it should define the possible forms in which such information might be disclosed and whether it must be marked "confidential." The definition may include the following:
  - Business/marketing plans and strategies
  - Financial information
  - Employee, contractor, customer lists
  - Business methods
  - Operating procedures
  - Pricing and sales data
  - Terms of commercial contracts
- **Exclusions from the Definition of Confidential Information** – There should be a section addressing items that are specifically excluded from the definition of confidential information, including, for example, information that becomes public by a means other than public disclosure by the recipient of the information.
- **Nondisclosure Obligations** – This section will typically impose a standard of care (e.g., best efforts, reasonable care) that the recipient must use to keep the information confidential.
- **Use and Access Restrictions** – Such restrictions will typically limit access to and use of the confidential information even within a recipient's own organization.
- **Safekeeping/Security Requirements** – An NDA may specify certain methods or procedures for the safeguarding of confidential information. You may be familiar with the use of a "data room" for such purposes.
- **Term** – NDAs may last indefinitely or, more commonly, NDAs will terminate on a specific date or upon occurrence of a specific event. The term often depends on the type of information involved.
- **(Optional) Non-Solicitation** – A non-solicitation provision prohibits one or both parties from soliciting or offering employment to the other party's employees.

- **(Optional) Non-Circumvention** – A non-circumvention provision is used to protect the ideas and opportunities under a business deal and, in the event the parties elect not to pursue a business relationship, it provides that neither party shall make any use of the other party's information.

NDAs come in two forms: mutual and unilateral. Mutual NDAs restrict both parties in their use of the confidential information provided, whereas unilateral agreements are used when only one party is restricted in using such confidential information (commonly because only one party has disclosed confidential information).

Tip: Parties should sign an NDA as early as possible in their relationship. If a party discloses information before signing the NDA, the NDA should specifically address coverage of prior disclosures.

10.1.2   Term sheets

A term sheet is a document that provides a summary of the key terms of a proposed transaction.

Term sheets surface in the initial stages of a transaction to show the intent of the parties to enter into an agreement, but they may not require the parties to finalize the deal on the exact terms in the term sheet. In essence, a term sheet is the modern-day handshake – it is a written confirmation of the fact that both parties have agreed to some terms in principle, but the deal remains subject to negotiation.

Key terms in a term sheet

Term sheets vary widely from deal to deal and there is no formula for drafting a term sheet. Nonetheless, here are some items to consider when drafting, reviewing or negotiating a term sheet:

- Description of the transaction and transaction structure
- Main deal terms
  - Purchase price
  - Description of assets/stock being acquired
  - Important closing conditions
  - List or description of ancillary agreements
- Important dates
- Scope/procedure/timeline of due diligence
- Confidentiality
- Exclusivity
- Restrictions regarding public announcement of the deal
- Tax issues

Term sheets are not used in every deal, however, they may be beneficial to help focus negotiations and enhance deal stability and commitment.

Tip: Generally, term sheets are non-binding, but they may contain some binding provisions. Binding provisions may include exclusivity and/or confidentiality (if not agreed to in a separate NDA, as discussed above). Term sheets should clearly identify both binding and non-binding provisions because, if there is ambiguity that leads to a dispute, courts may interpret the whole term sheet to be binding based on the intent and actions of the parties.

10.2 BUSINESS/DEAL STRUCTURING DECISIONS

10.2.1 Choice of entity

One of the first considerations in starting a new business (or forming a project company) is determining the form the business will take, which generally requires considering whether the business will be incorporated or take some other entity form. Tax factors are often the driver of the entity formation decision, but there are a number of non-tax factors that should be carefully considered as well.

Sole Proprietorship. A sole proprietorship is an active business carried on by one individual: its sole proprietor. It is the simplest of all business entities to organize. A sole proprietor is ideal for someone who wants to be in a one-person business, have little start-up cost, have little or no liability, and does not seek investment from outsiders.

Tip: A sole proprietorship is strongly advised against as a means for project development, as sole proprietors have unlimited personal liability, and the personal assets of proprietors can be attached to satisfy claims against the business.

General Partnership. A general partnership is an association of two or more individuals who conduct a business as co-owners. Its two outstanding features are the unlimited liability of each partner for all the debts of the business, and the implied authority of each partner to bind the firm to outsiders by any act within the scope of the usual and ordinary activities of the particular business.

Tip: General partners in a general partnership also bear unlimited personal liability; accordingly, this entity form is ill advised for solar project development and finance.

Limited Partnership. As with a general partnership, a limited partnership is an association of two or more individuals who want to conduct a business as co-owners. The distinct difference, however, is that there are two kinds of partners, general and limited. General partners have unlimited liability and manage the business. The limited partners' liability is limited to that of their investment, as long as they do not take management roles within the company.

Tip: There are two main disadvantages to forming a limited partnership: the general partner bears unlimited liability, and unanimous consent is generally required to sell partnership interest and admit new members.

C Corporation. A C corporation is a legal entity with a legal existence that is separate from its investors and shareholders. The net profits of the company could be subject to double taxation, once at the entity level, and again on dividends distributed to shareholders.

Tip: C Corporations are generally favored due to the liability protection that they offer; however, the double-taxation inherent in forming a C Corporation is a deal-breaker for most people.

S Corporation. An S corporation is a type of corporation with a special election filed with the IRS under Sub-Chapter S of the Internal Revenue Code.

An S corporation has the legal advantages of a C corporation, with the additional benefits of pass-through income/loss, and is not subject to double taxation, unlike a C corporation.

Tip: While S Corporations generally provide the same benefits as limited liability companies (discussed in Table 10.1), they generally entail corporate formalities similar to

Table 10.1 Choice of entity issues.

Issue	Limited partnership	S Corporation	C Corporation	LLC
Definition	Association of 2 or more as co-owners. General partners have a voice and limited partners have no voice.	Legal entity made up of centralized management and Board of Directors. Separate entity from its shareholders.	Legal entity made up of centralized management and Board of Directors. Separate entity from its shareholders.	Either member-managed or have separate management.
Limited Liability	Limited liability only for limited partners who do not take too active a role in management; general partner has unlimited liability.	Limited liability for shareholders even if they participate in management.	Limited liability for shareholders even if they participate in management.	Limited liability for members even if they participate in management.
Management	General partner makes the decisions. Limited partners do not have a role in management.	Board of Directors and officers make all decisions; shareholders allowed to vote on fundamental changes in corporation only. Shareholders elect and remove board members, but can't act on behalf of the corporation or participate in the daily management.	Board of Directors and officers make all decisions; shareholders allowed to vote on fundamental changes in corporation only. Shareholders elect and remove board members, but can't act on behalf of the corporation or participate in the daily management.	Operating agreement provides who shall manage. If no agreement, then management is determined by share of contribution.
Pass-Through Tax Treatment	Yes	Yes	No	Yes
Management	By general partner.	By Board of Directors (who may be shareholders)	By Board of Directors (who may be shareholders).	By all members, unless manager(s) appointed.
Operational Formalities	Few	Corporate formalities must be observed.	Corporate formalities must be observed.	Few
Financing	New partners, loans, injections from existing partners.	Issued stock, loans, selling of equity (bonds).	Issued stock, loans, selling of equity (bonds).	Selling of membership interests.

those of a C Corporation, a fact which often makes a limited liability company (LLC) the entity form of choice.

Limited Liability Company. A limited liability company is a relatively new corporate entity that was designed to combine the benefits of partnerships and corporations – i.e., for owner liability purposes, it resembles a corporation, and for federal tax purposes, it resembles a partnership. Thus, a limited liability company is ideal for individuals who seek the flexibility and pass-through income of a partnership, coupled with the liability protection afforded by corporations.

Bottom Line: LLCs are usually the entity structure of choice for solar project developers, as they provide limited liability and pass-through taxation along with flexibility in respect to corporate formalities.

10.2.2   A holding company versus a project company

As mentioned above, the limited liability company is the most common entity of choice for project companies of solar project developers. Furthermore, project developers will often employ an entity structure whereby there is a parent company (or holding company) that owns the membership interests of one or more project companies. Most developers will form a project company for each solar project that they develop. There are numerous reasons for employing this structure, including facilitating future transfer or sale, and project financing of the project. For project financing purposes, utilizing a project company structure is key to isolating the assets and liabilities of specific projects. This becomes particularly important in non-recourse or limited recourse project financings.

10.2.3   Where to form your business

Corporations and LLCs are often formed in one of two places: Delaware or the state in which they are doing business. Delaware is often considered the formation state of choice because of its perceived favorable laws governing corporations and LLCs, as well as its low fees and annual taxes. Furthermore, Delaware has a well-developed body of law governing corporations and LLCs because so many of them are formed in the state. On the other hand, if you conduct business in only one state, it is typically cheaper and often easier to form your entity in that state.

Tip: Corporations and LLCs must register to do business in each state in which they operate, and there are filing requirements and fees associated with registration. An entity is considered to be a "foreign corporation" or "foreign LLC" in any state other than its state of incorporation or formation.

10.2.4   Steps to create an LLC in Delaware

Table 10.2 summarizes the primary steps in forming an LLC in Delaware.

Tip: Most developers create an LLC for flexibility. Typically, LLCs are formed in Delaware, but developers may opt to form the LLC in the state where the project is located.

Table 10.2 Steps for delaware incorporation.

Step		Description
1	Reserve a Unique Business Name	The company should select a name for the new business. The name must not be used by another business in Delaware. To ensure that the name has not been taken, the company may search the State of Delaware's database of existing business names at http://sos-res.state.de.us/tin/EntitySearch.jsp. If the company chooses to form an LLC, Delaware requires that the business name include either the acronym "LLC" or the words "limited liability company." Once the company has agreed upon a name, the name may be reserved by filling out a form available at *http://corp.delaware.gov/llc-nameres.pdf.*
2	Obtain a Registered Agent	All new business entities must obtain a registered agent, either an individual or a business entity, for the service of legal documents. This registered agent must already be authorized to practice business in Delaware, and have a physical address within the state of Delaware.
3	File a Certificate of Formation	The company must file the appropriate "Certificate of Formation" for the appropriate type of business entity with the Delaware Division of Corporations.
4	Create the Organizational Resolutions & The Resignation of the Organizer	On the effective date of the Certificate of Formation, the organizers of the company should create the Organizational Resolutions and the Resignation of the Organizer. Once these forms are completed, the initial member of the LLC replaces the organizer of the LLC.
5	Draft & Receive Written Consent from the Members of the LLC	On the effective date of the Certificate of Formation, the LLC should draft and receive written consent from the member(s) to approve the Certificate of Formation, elect managers, adopt an Operating Agreement, and ratify past acts. Before electing members, it should be noted that members of an LLC may be personally liable if they breach the duty of loyalty.
6	Draft an Operating Agreement	On the date of filing the Article of Incorporation, the members should create an Operating Agreement.
7	Draft & Receive Written Consent of the Members of the LLC	On the date that a company files the Articles of Incorporation, the members should receive written consent of the members to adopt the Operating Agreement, appoint managers and/or officers, and ratify past acts.

10.2.5   Joint ventures

Joint ventures are another type of entity or ownership structure. In general terms, a joint venture is an arrangement in which two or more parties combine to achieve a common purpose or goal. A joint venture can take many different forms, from simply a contractual agreement between parties to the formal formation of a joint venture entity. There is no "one size fits all" when it comes to joint ventures; however, the following is a summary of various reasons parties undertake joint ventures and several common issues to consider when entering a joint venture.

Reasons for a joint venture

- **New Market Entrant; Strong Local Player.** One of the primary reasons for joint ventures or similar arrangements is that it may allow a party to quickly and efficiently enter a new market. Often, a developer may require expertise in terms of developers, engineers, financiers, etc. to develop, construct, finance and operate a project, but may lack local knowledge and contacts that may be necessary to secure or finish a project. A joint venture with a local party containing such information may be a convenient structure to bridge that gap. In addition, entering a new market through a joint venture, at least initially, is generally quicker and cheaper than doing so alone.
- **Expand and Secure Additional Business.** Often, a joint venture is an efficient structure for parties to secure and expand their pipeline of projects. This is true not only for developers, but also for manufacturers and contractors. Because utility scale energy projects usually take several years to develop, developers often need fairly accurate and firm information regarding supply and construction costs early in the process, typically well before they are able to make firm purchases or contract commitments. Manufactures and contractors are generally not in a position to make firm commitments so far in advance of the actual purchases. Manufacturers and suppliers, however, are often very interested in securing an early position for the supply and construction of projects, and, as a result, a joint venture is often a beneficial structure that the parties can use to give a developer and suppliers and contractors more certainty regarding supply and construction.
- **Risk sharing.** A joint venture may be a simple mechanism for reducing and sharing risks. As noted earlier, this is particularly true with long-term resource commitments, and in such instances, may be a fair and convenient structure for sharing and allocating related risks.
- **Size of Project.** More and more, utility scale energy projects are growing, with project costs commonly totaling hundreds of millions of dollars; in some cases exceeding a billion dollars. The financial burden and responsibility of such a project is often too large for a developer to shoulder alone, and thus a joint venture with other parties is beneficial.
- **Allocation of Tax Benefits.** Projects rely on, and are the beneficiaries of, significant tax benefits from tax credits to depreciation, many of which are only allocable to the owners of the facility. As a result, joint venture structures are commonly employed to maximize and efficiently allocate tax benefits to the parties best able to utilize them.
- **Restrictive Covenants.** In circumstances where a developer or potential investor is subject to restrictive covenants under loan agreements or other contracts or documents, a joint venture is a structure that may allow the party to avoid the restrictive covenants and participate in a project.

Potential downsides of a joint venture

- **Different Corporate Objectives, Strategies and Desires.** An important exercise for parties prior to entering into a joint venture is to assess and determine their respective internal objectives and desires, as well as those of the other parties, to determine whether the proposed joint venture is a good fit, both in scope and duration. Often, a mismatch in objectives, strategies or desires of the parties results in friction and paralysis in joint ventures causing more harm than good.

- **Cultural Differences.** Similarly, a difference in corporate culture may result in a dysfunctional joint venture. For example, one party may be a fast-paced, quick-to-act company and the other is a methodical, bureaucratic organization with several layers of approvals required for decisions. A joint venture between the two parties may not be a good fit without the two first being aware of these differences and developing a structure and protocol that manages expectations and provides flexibility for each party to conduct business in a manner that fits into its own business culture.
- **Lack of Trust.** Oftentimes joint venture parties are competitors or adverse parties in other transactions. In addition, conflicts of interest of the parties are inherent in joint ventures. As a result, it is not uncommon for there to be a lack of trust between parties, which if not managed properly, can render the joint ventures ineffective.

Common issues to consider

- **Due Diligence.** As described above, there are various reasons for undertaking a joint venture and various potential downsides. As a result, it is important for parties to take time to assess the proposed arrangement, including the compatibility of the parties and their goals.
- **Management and Control.** Generally, management of a joint venture is vested in a board of directors, with directors appointed from the various parties depending on the respective ownership of the parties. Parties may also seek to have control in the day-to-day management of the joint venture, which typically is vested in officers of the entity. As a result, it is common for parties to have the right to appoint certain key officers or for the parties to agree that they will hire third parties to perform the responsibilities so that neither of the joint venture parties has all of the control. Parties may also establish appropriate and regular reporting protocol so the parties are kept up-to-date on the development and activities of the joint venture.
- **Liability.** In general, most joint venture structures are set up to limit the liability of the parties, creating nonrecourse structures, generally limiting the liability of the members to the investment in the venture. Joint venture members may also further attempt to limit their liability by creating special purpose entities as their contracting parties. One area of potential liability exposure to consider is that of the officers and directors. Typically, officers and directors will owe certain fiduciary duties, including duties of care and loyalty, and depending on the jurisdiction, may have additional obligations.
- **Noncompetition.** For many arrangements, it may be important that the parties not compete with the joint venture during the term of the joint venture to properly allow the joint venture time to develop and to reduce part of the inherent conflict of interest of the parties. In addition, it may be appropriate for there to be a cooling off period after the term of the joint venture under certain circumstances. The parties will need to discuss the appropriate length and scope of such noncompetition restrictions.

- **Transfers.** Joint ventures will generally have various restrictions on transfers of interest, in large part, because the parties typically enter the joint venture because of the particular counterparty. However, in many circumstances, transfers may be necessary and appropriate. The parties should discuss and consider what circumstances are appropriate for them, including the termination and exit strategies.

Termination: exit strategies

Joint venture partners rarely enter a joint venture with the intention or expectation that the joint venture will fail, and as a result are often reluctant to discuss in any detail termination or exit strategies. Regardless of the expectations of the parties, one of the key provisions of a joint venture is the termination provision. It is critical that the parties consider and discuss what termination events are appropriate and what the procedures will be, including deadlocks, failure to agree to budgets or capital contributions, breach or insolvency of a party, etc. A properly structured termination provision may also be the mechanism that actually keeps the parties honest and the joint venture successful.

10.2.6   Acquiring a project in development

Many project developers choose to acquire a project that is already under development. Projects may be acquired at any stage of development. Developers often acquire projects that are construction-ready (or close to being construction-ready), meaning that site control and all permits have been obtained. In such an instance, there will need to be some type of formal acquisition. Any developer considering such an acquisition should consult with their attorney, but in general there are three different types of acquisition structures.

1) *Stock Purchase.* In a stock purchase, the Buyer (or the Buyer's holding company) purchases the outstanding capital stock of the project company (the "Target") directly from the Target's shareholders. In a solar project acquisition, the Target's shareholders typically will be few in number. A stock purchase in many ways is the simplest form of acquisition and is the most prevalent.
2) *Asset Purchase.* In an asset purchase, the Buyer acquires all or selected assets of the Target and assumes all, a portion, or none of the liabilities of the Target pursuant to an asset purchase agreement. From a Buyer's perspective, an asset deal is generally the safest and cleanest approach because the Buyer only assumes the agreed upon liabilities. However, in the solar energy space, because most projects are held in special purpose entities with no other business than the project, the potential exposure to the Buyer is generally lesser. In addition, asset deals often require additional third party consents to transfer assets. As a result, asset deals are less common in solar transactions.
3) *Merger.* While mergers are one form of acquisition structure, they generally are not used to acquire rights to a solar project.

Tip: Two important due diligence points in a solar project acquisition to keep in mind: Buyers should confirm which entity holds or was issued permits (generally you want it to be in the project company's name), and Buyers should review any assignment or change of control restrictions on any permits or other agreements in the project company's name.

10.3   PROJECT DEVELOPMENT CONSIDERATIONS

10.3.1   Site control

Apart from the outright purchase of a site, there are generally three different ways to achieve site control: easement, lease or option. An easement is an interest in land in the possession of another that entitles the holder of the interest to a limited use or enjoyment of the land. A lease is a conveyance of land for a term that is less than the term of the owner's interest. An option provides for the exclusive right to lease property or obtain certain easement rights at some time in the future.

Easement. An easement may be either exclusive or non-exclusive. Under an exclusive easement the easement rights are exclusive to the holder. If easement is nonexclusive others could have the right to use the land concurrently. It should be noted that the grantor of an easement relinquishes very few rights even if the easement is exclusive. Easements for a short period of time are sometimes referred to as temporary easements. The rights granted under an easement may be permanent or for a set period of time. Finally, payment for an easement can either be in a lump sum or through periodic payments.

Lease. A lease creates a landlord/tenant relationship. The tenant usually has exclusive use of the property covered by the lease. If the landlord wants to retain rights to use land, such rights must be specifically stated. Unlike an easement, a lease is always for a set period of time. Lease payments are usually periodic.

Tip: In addition to the notes above, easements and leases differ in subtle ways, including treatment in bankruptcy court and tax foreclosure. Also, in some states, leases for a term greater than 35 years (including extensions) may be considered a conveyance.

Option. An option provides for the exclusive right to lease property or obtain certain easement rights at some time in the future. The right must be exercised within a specified period of time. The time period can vary from several months to several years (3–5 years is common). The option provides for some form of payment for the option rights being preserved. Most importantly, an option typically allows the developer to walk away for any reason within the option period.

10.3.2   Power purchase agreements

To successfully structure project finance for a facility, generally there must be an assured market for the power produced for the duration of any third-party financing provided. The price of the energy sold by a project must be sufficient to generate adequate revenues to cover debt obligations and expected operating expenses. Thus, the power purchase agreement ("PPA") is vital to project financing. It is important that developers evaluate and negotiate the PPA from a lender's point of view.

Project financing

Solar projects are overwhelmingly project financed. In a project financing, the lender looks to the future cash flows of the project – for a solar project, the revenues under the PPA (and associated tax credits) – for repayment. Lenders typically have a lower risk tolerance than developers and equity investors. As a result, the PPA is the central and most important document in a solar project. It is the revenue driver, and most other documents will be molded around the PPA.

Developers often negotiate and settle a PPA before the project lenders are selected. As a result, the developer must attempt to anticipate and address lender requirements in the course of negotiating the PPA. A failure to do so may result in the renegotiation of the PPA as a condition of the lender providing financing, and such renegotiation should be avoided.

Pricing

The first and most important issue for the PPA is pricing sufficiency. Often, to sign a PPA, a developer will shave the price as low as possible. Unfortunately, this is sometimes overdone. The pricing must be sufficient to accommodate project costs, including financing costs, and transaction costs: interest, guaranteed returns, bank fees, legal fees, consultant fees, as well as actual project operation and maintenance cost. The lender will run its own projections based on very pessimistic views of both project performance and operating costs, and the revenue must be sufficient to survive this pessimism.

Performance standards and liquidated damages

A second issue is performance standards and liquidated damages. This is a complex issue. The optimal strategy is to create terms that are predictable and pass-through to others.

The damages under a PPA must be predictable. Some utilities want performance shortfall to result in indemnification for regulatory fines imposed on them for failing to meet their RPS obligations. These provisions are dangerous, as in most states the regulations do not provide for specific fines, but simply fines to be determined by the regulatory agency. With no guidelines to help banks place a value on possible fines, they are sure to pick a very large number for their projections. Similarly, the PPA may allow for replacement energy or replacement Renewable Energy Credits (RECs); this is only helpful if there is a liquid market with predictable pricing for the replacement energy or RECs that meet the requirements of the PPA. Developers should not let liquidated damages or replacement energy/RECs become a black hole of uncertainty. The potential financial downside of underperformance must be predictably quantifiable.

Liquidated damages and other performance-related costs should be subject to pass-through to others whenever possible. If the PPA has an availability guarantee, a developer should limit that guarantee (and associated damages) to the availability guarantee provided by the equipment supplier. In this fashion, performance shortfall risk is taken by the equipment supplier rather than the project. Even if the PPA has a generation guarantee instead of an availability guarantee, a developer should try to work the required generation levels around the equipment supplier availability guarantee equivalents, so that the risk can be passed through. Beyond the equipment suppliers, a developer should try to negotiate the PPA performance requirements and penalties to take advantage of any other guarantees from third parties, as well as other project factors. If a developer has reason to believe that its project will be facing some downtime for equipment replacement, for instance, it must make sure that the PPA performance requirements accommodate this, and that the equipment warranty cover any resulting PPA losses. There is a small army of contractors and suppliers behind the project owner: EPC contractors, equipment suppliers, O&M contractors – there is no need for the project to retain more performance risk than absolutely necessary.

Bundled or unbundled PPAs

Renewable energy consists of two distinct commodities that may be sold together or separately. These two commodities are power and environmental attributes. The environmental attributes are associated with the production of renewable or "green" energy and may also be known as renewable energy certificates, renewable energy credits or solar renewable energy certificates. These environmental attributes are typically referred to as "RECs," which typically represent the environmental attributes from one megawatt hour of electricity produced by a renewable energy source.

Environmental attributes may be an important revenue stream for a solar project. Thus, it is important to note whether a PPA includes or excludes the sale of RECs with the power. A "bundled" PPA is one in which the seller is selling both the power and the environmental attributes, while an "unbundled" PPA includes only the power. In an unbundled PPA, the developer may sell the RECs to a different purchaser under a REC contract.

Other key terms and important questions to ask when reviewing a PPA

A typical PPA may appear to be overloaded with legalese. Nonetheless, even a first-time reader can quickly understand the most important deal points by asking the right questions. Set forth in Table 10.3 is a short guide that identifies key terms/concepts to spot in a PPA, along with important questions to ask.

10.4 CONSTRUCTION DOCUMENTS

As mentioned earlier, financing is a central challenge most developers of solar energy projects face. Therefore, it is very important that a prospective project be free of flaws, and that the project documents are pristine for lender review in a project financing.

Financeability

Project developers can satisfy this scrutinized lender review by keeping in mind that the project cash flow needs to be predictable and uninterruptible. When lenders (and their counsel) review project documents, they look for anything that could interrupt cash flow or make cash flow less predictable. In order to obtain necessary credit approvals, the lenders need to be able to make reliable financial projections for the

Table 10.3 Key terms and questions for reviewing a PPA.

Key term	Question to ask	Notes
Term	• What is the duration of the PPA?	A typical PPA term is 20–25 years.
Price	• Does the price stay the same for the duration of the PPA? If the price increases, when? If so, by what amount does it increase?	A typical PPA will escalate annually at a rate of 1–3%.
Commercial Operation Date (COD)	• When does COD begin under the PPA? • Must the Purchaser certify that COD has occurred or is the developer simply able to notify the Purchaser? • Can COD occur in phases or must it be all at once? • Is there a guaranteed COD date, and if so, what are the purchaser's remedies (Liquidated Damages, termination)?	
Purchaser Obligation/ Output Guaranty	• Is the purchaser obligated to purchase 100% of system output? • Is there an amount of output that is guaranteed by the seller? • What damages are associated with a failure to produce such output? • Is there a lower price for energy produced that exceeds the guaranteed output? • Does the EPC contract or any supply agreement back up the level of output and damages under the PPA?	Since the output from solar projects is generally predictable and solar projects can be sized to meet a purchaser's needs, PPAs for solar projects (and others) typically require the purchaser to purchase 100% of the output.
Availability Guaranty	• Does an availability guaranty exist, whereby the panels will be available a certain percentage of the time, excluding hours lost to force majeure and a certain amount of scheduled maintenance? • If so, over what time periods are such guaranty measured?	
Curtailment	• What rights, if any, does the purchaser have to curtail output? • Is the seller reimbursed for curtailed energy?	The purchaser may have some rights to curtail energy (e.g., in emergency situations), but these rights should be capped.
Commitment to Develop; Security	• Does the PPA require a security deposit from the seller? • When is the deposit required (execution or COD) and when is it returned?	A performance security may take the form of a parent guaranty, letter of credit or performance bond.
Milestones and Delay Damages	• Does the PPA contain milestones? If so, are they realistic? • What remedies are required by the developer for failure to meet a milestone? Does the purchaser have the right to terminate the PPA, collect delay damages, or require the seller to post additional credit support?	
Liquidated Damages (LD)	• Are there performance LDs? • What is the formula for calculating LDs? • Is there a cap on LDs (annually or aggregate basis)?	

Default	• Determine what events constitute events of default, which may include:
	  • failure by any party to pay an amount when due
	  • other types of material defaults
	  • bankruptcy, reorganization, liquidation or similar proceeding of any party
	  • a material default by a party's guarantor
	• Is there a cure period for a default?	

Title, Risk of Loss	• Where do title and risk of loss pass? • Where is the delivery point?	
Termination	• Determine the "off-ramps," if any, that allow one or both of the parties to terminate the PPA prior to COD.	
Lender Protection; Assignment Cooperation	• Confirm whether the PPA has customary lender protections (the purchaser will cooperate, notice to lender, consent to collateral assignment, etc.)	

life of the financing. For those projections to be meaningful, the lenders must have confidence that the cash flow will continue uninterrupted and stay within the predicted range. If the lenders cannot gain this confidence then they cannot create reliable financial projections, and they will not provide the financing.

Once a project developer understands the lender's motivations (predictability and uninterruptibility), it becomes a relatively simple matter to review the documents and identify provisions that could interfere with these motivations. Lenders notoriously do not like to take risks – they generally do not distinguish between "unacceptable risk" and "general risk." Therefore, when reviewing project documents, specifically construction documents, it is important to simply look for any possible risks.

Typical construction documents

The various construction documents required for a solar project are critical to the project's financeability. Accordingly, the developer of a solar project must enter into agreements for the following:

- Design and engineering
- Procurement of solar modules, inverters and mounting systems, if necessary
- Obtainment of construction services necessary to install equipment
- Operation and maintenance of the completed facility

Engineering, procurement, and construction tasks are often combined in a single agreement called an "EPC Agreement." There may be separate agreements that provide for or anticipate other services, including warranty services.

Alternatively, all phases of the design and engineering, procurement, and construction/installation services are sometimes addressed in a single agreement ("turnkey" agreement) and a single entity is made responsible for the whole project. It is also common to have separate agreements such as design and engineering agreements, construction/installation agreements ("balance-of-plant agreements"), and procurement and sale agreements for major pieces of equipment, using one or more contractors for each of the various services. Depending on the contractual structure, warranties, insurance, and other matters may be addressed in a single master agreement or in individual agreements. Whatever the contractual structure, there are some key provisions that are critical to financeability that should be evaluated in any construction document.

Evaluating a construction document

Table 10.4 presents some of the deal points for the major construction documents associated with solar projects. There is no single determinant of financeability. Therefore, few of these issues are fatal by themselves, unless the violation is egregious. Instead, financeability is predicated on the totality of the risk package. Nevertheless, every small risk that is added to the package makes it more likely that the project will not pass investor/lender scrutiny, so no risk should be casually accepted.

Table 10.4 Key terms for construction documents.

Key term	Note
Scope of Work	• Determine whether the contract is "turnkey" or for limited services only • Determine what scope is excluded
Permits	• Determine which permits the contractor will obtain • Developer's permit obligations to be listed
Payment Schedule	• Can be either milestone based or on a monthly basis • Monthly typically includes 5–10% retainage • Determine whether the contract provides that achievement of milestone is verified by an independent engineer
Taxes	• Determine whether sales taxes are included in the contract price
Cancellation Fees	• Determine the cancellation fees, if any
Change Orders	• Determine procedure for requesting a change in the order
Subcontractors	• Determine whether use of subcontractors requires the approval of the developer for work over a certain cash threshold ($100,000 is a typical threshold)
Dispute Resolution	• Determine whether there is a dispute resolution clause and to what extent it limits dispute resolution options and whether it precludes the possibility of taking a matter to trial
Work During Dispute	• Check for provisions requiring the contractor to work during a dispute with developer
Guaranteed Completion Dates	• Assess whether agreement provides guaranteed dates for substantial/final completion
Event of Default	• The developer should have the right to terminate the contract and assume control of the project if there has been a material breach that remains uncured after a specified period of time
Delay Liquidated Damages (LDs)	• Are there delay damages to compensate for: • Financing costs • Damages under a PPA or SREC Agreement • Damages for lost tax benefits
Performance Testing/ Guarantee	• Performance testing to occur after substantial completion and, possibly, annually thereafter • Determine whether there is an output guarantee
Bonuses	• Determine if there are any bonuses payable in the event of early completion or over performance of the system. These are somewhat uncommon
Liability Caps	• Check the liability caps, if any
Warranties	• Ensure contractor warranties its work and assigns all equipment warranties to the project company at the expiration of the warranty period
Performance Security	• Determine whether the contractor is providing performance security during the construction period and warranty period
Lien Releases	• When developer makes periodic payments, it should obtain lien releases • Confirm that the form of lien release complies with local law
Tax Credits/ Incentives	• The contract should require cooperation from both parties in submitting all required documentation for any federal, state or local incentives • Determine who bears the risk if the incentive is not received • Determine whether either party in the contract is making any representations or warranties about receiving incentives
Lender Protection; Assignment Cooperation	• Confirm whether the agreement has customary lender protections (e.g., cooperate, notice to lender, consent to collateral assignment) • Confirm whether assignment to a lender requires approval

Analysis of deal points

As mentioned previously, the financeability of a construction document depends on the totality of the risk package. In conjunction with the checklist above, five specific provisions that are material to financeability should be mentioned:

1. Equipment warranties

Equipment warranties likely will be subject to substantial negotiation. The issues to carefully consider when negotiating an equipment warranty include the following:

1) the term of a particular warranty;
2) whether the term of the warranty can be extended;
3) the definition of a "defect" with respect to a piece of equipment;
4) any limitations on a warranty, including limitations related to acts of third parties (e.g., O&M providers); and
5) the remedial measures a contractor may take to cure a defect.

In addition to these points, another contract-drafting consideration is the extent of the warranty – whether the contractor will obtain "commercially reasonable" warranties or the "best available" warranties. Finally, construction documents should address whether the contractor will "pass-through" warranties received from its suppliers and subcontractors.

2. Performance guarantees

Project financing is much easier to acquire when there is a performance guarantee from contractors and/or equipment suppliers in place. A performance guarantee provides certainty, which enhances a project's financeability. This gives the financier comfort that the solar project will produce a baseline level of output or otherwise receive a payment in lieu of any output (and be assured of at least a certain revenue stream). Thus, from a developer's perspective, it is critical to have a performance guarantee in construction documents.

3. Liquidated damages

Liquidated damages may be another area of extensive negotiation. There are two general types of liquidated damages: performance liquidated damages and delay liquidated damages.

Performance liquidated damages are assessed when a project falls short of its guaranteed performance. Accordingly, the performance liquidated damages will be calculated pursuant to a formula specified in the construction documents, which is rooted in compensating the developer for a shortfall in production from the project.

Delay liquidated damages are relevant when a project misses its deadline for any guaranteed dates of completion and/or other milestones. They are designed to compensate the project owner for the revenue lost as a result of such delay. Thus, delay liquidated damages come in the form of a per day assessment for each day the project has missed a guaranteed deadline; some delay liquidated damages incrementally increase at certain thresholds (often in 10- or 15-day increments). Delay liquidated damages may be subject to a cap within the limitation of liability provision in a construction document. In sum, delay liquidated damages enhance financeability because they provide assurance that, in the event a project does not start on time, the developer will still receive revenue that otherwise approximates the amount of revenue it would have generated but for the delay.

4.   Limitation of liability

Contractors and suppliers often seek to limit their liability under construction contracts. A contract may include a general limitation of liability. For example, construction contracts often limit the liability of the contractor to the contract price. Additionally, construction documents also may have "sub-caps" to limit liability for specific items. For example, a construction contract may limit the liability for liquidated damages to 10% of the contract price. This liquidated damages sub-cap may be further broken down to differentiate between performance liquidated damages and delay liquidated damages. These limitations and sub-caps are often heavily negotiated between parties and are of great interest to financiers. In sum, the limitation of liability helps financiers evaluate the downside risk for a project, and this is another important component in obtaining financing.

5.   Performance security

Construction documents often specify a certain type of security provided by the contractor to the developer. Performance securities can come in several varieties: bond, letter of credit, or parent guaranty. Such security is meant to ensure:

1)   the timely performance of the contractor;
2)   that such performance on the project is completed pursuant to the construction documents; and
3)   that no liens or any other encumbrances are filed against the project property or improvements.

In addition, albeit rare, contractors may demand some form of reciprocal security issued by the developer to ensure prompt and full payment of all the developer's obligations under the construction contracts. In negotiating the performance security provision, contractors will also request an opportunity to cure any default or delay and will try to limit a developer's ability to call on the contractor's performance security. The performance security is just one more assurance that financiers look for when evaluating the downside risk of a project.

10.5   CONCLUSION

The central principle when evaluating construction documents is this: ALL documents must pass muster, individually and collectively, not just the "important" documents. As a result, project developers should be prepared to present a consistent and cogent set of construction documents to lenders and/or investors. Additionally, project developers should be prepared for the possibility that lenders and/or investors will require the developer to make substantial changes in the construction documents in order to provide reasonable assurances of the revenue flow. Ultimately, financiers want to ensure predictability and uninterruptability of the cash flows.

Chapter 11

Business: Finance

Michael Mendelsohn

Financing is a critical component to renewable energy deployment. Accessing financial capital can be a difficult barrier that can slow or stop a wide range of renewable energy projects, but is most burdensome to projects that employ relatively new technologies, are under development by less experienced entities, or are under contract with utilities or end-users (also referred to as off-takers) that have lower credit ratings or cannot ensure full access to relevant transmission infrastructure (Mendelsohn & Harper, 2012).

Even for projects in which capital is available, it may be at rates or yield requirements that are burdensome to project economics. Financing costs can significantly raise the final cost of power. In Figure 11.1, the financial carrying costs – including the equity and debt components – can be larger than the recovery of initial capital expenditures (i.e., depreciation). In this chapter we will define the basics of project finance for renewable energy projects. We will also give specific emphasis to financing mechanisms used in the market at present as well as highlight potential areas for improvement in renewable energy finance.

The specific impact of financing on the cost of renewable energy is related to a wide variety of factors including the initial cost of the system, the natural resource available (e.g., sunshine), the design, operation, and maintenance of the system.

In Figure 11.2, a hypothetical solar system in Phoenix is compared under three different financing scenarios – no financing cost, a relatively low financing cost (represented by debt with a 6% yield requirement and equity with a 10% after-tax return), and a high-cost financing scenario (represented by debt with 10% interest rate and equity with 18% after-tax return expectation). In short, even the low-cost financing scenario approximately doubles the Levelized Cost of Electricity (LCOE) requirement over the no-cost (or, outright purchase) scenario. The high-cost scenario represents an approximate tripling (or 200% increase) over the no-cost financing scenario. All cases assume $1/watt installed cost consistent with the Department of Energy's SunShot initiative goals (Wesoff, 2012).

According to Deutsche Bank Climate Change Advisors (DBCCA), every 1% reduction in target equity internal rate of return (IRR) results in $4/MWh reduction in wind and $8/MWh for PV (DBCCA, 2011).

11.1 FORMS OF INVESTMENT

Financial capital generally comes in three primary types – debt, "mezzanine" capital, and equity. Debt can be sourced from banks or public sources. In 2011, MidAmerican Energy raised nearly $1.3 billion in two separate debt financings to support

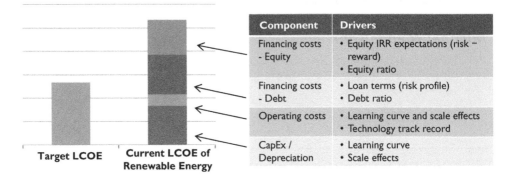

Component	Drivers
Financing costs - Equity	• Equity IRR expectations (risk – reward) • Equity ratio
Financing costs - Debt	• Loan terms (risk profile) • Debt ratio
Operating costs	• Learning curve and scale effects • Technology track record
CapEx / Depreciation	• Learning curve • Scale effects

Target LCOE **Current LCOE of Renewable Energy**

Figure 11.1 Components to Levelized Cost of Electricity (LCOE) (Deutsche Bank Climate Change Advisors (DBCCA), "Get fit plus: Derisking clean energy business models in a developing country context". 2011).

the development of the Topaz solar facility without government guarantees or other credit enhancements. The event was heralded as a watershed moment as it indicated a maturing of the industry and the appeal of renewable energy facilities among the investment community (Businessweek 2012).

Mezzanine finance refers to financial capital that may be less secure than debt but less risky than pure equity ownership. In the U.S., "tax equity", a mezzanine finance product, is frequently used to monetize tax benefits (investment or production tax credits and accelerated depreciation) (Mendelsohn & Harper, 2012). Limitations of tax-based policies to induce investment are discussed later. Equity is generally provided by developers throughout the development cycle before tax equity is raised.

Debt is the lowest-cost source of capital, and generally prized by developers as it reduces the LCOE or the required price under a long-term contract (referred to as a power purchase agreement, or PPA). But debt is not easily obtained, particularly if the market perceives technology risk or other risk factors. Lenders to relatively riskier projects will require one of three components of compensation: higher yield, higher debt service coverage ratios (DSCR), or shorter term or duration of the debt (also referred to as the debt tenor).

DSCR represents the ability of a project or firm to pay the debt service (principal and interest) payments. The DSCR is calculated as the ratio of operating income (also known as Earnings Before Interest, Taxes, Depreciation, and Amortization, or EBITDA) over the debt service payment. A higher DSCR required by the lender indicates that the lender perceives more risk in the project's cash flows, and in turn, will limit the size of the loan to ensure it can be repaid even if production is below expected levels, operating costs are higher than expected, or contingencies occur.

Figure 11.3 provides DSCR and debt term data for different renewable energy technologies. Shorter debt terms may indicate higher risk perception potentially due to perceived technology risk, concern over long-term operational and maintenance expenditures, or simply because the overall debt arrangement (combination of yield, duration and other criteria such as required reserves) provides the best return for the developer and/or tax equity investor.

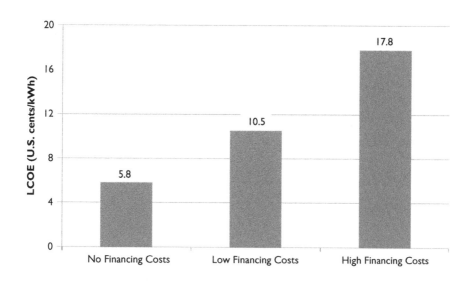

	No Financing Costs	Low Financing Costs	High Financing Costs
Debt	None	50% at 6% interest	40% at 10% interest
Equity (after tax)	None	50% at 10% Return on Equity (ROE)	60% at 18% ROE

Figure 11.2 Levelized cost energy from a PV System under 3 financing scenarios using NREL's SAM model (Mendelsohn 2011a).

11.2 FINANCIAL STRUCTURE

How projects incorporate the different financial components (debt, mezzanine, equity) is based on the availability and cost of those components, the allocation of risk and reward different providers of capital are willing to accept, and the policies that support capital investment, power production, or other facets of renewable energy project deployment.

In the U.S., renewable energy policy largely relies on tax benefits designed to promote private sector investment in renewable energy power projects (Mendelsohn & Harper, 2012). First, investors in renewable power generation projects can claim credits against their income tax obligations that include:

1) Investment tax credit (ITC), which is currently 30% of eligible project capital costs for solar and certain other renewable technologies.
2) Production tax credit (PTC), which currently ranges – depending on the technology – from $0.011 (e.g., for geothermal) to $0.022 per kWh (e.g., for wind). Opportunities to convert the PTC into other support structures are discussed below in section 11.6.1 (U.S. Policy to Induce Renewable Energy Finance). (Mendelsohn and Harper, 2012.)

Table 11.1 Classifications of financial capital (Mendelsohn, 2011b).

Investor class	Risk tolerance	Metrics	U.S. Range	Sources
Debt	Low	Debt service coverage ratio (DSCR), Interest Rate	5–9% (market)	Bank debt, Private placements, Public markets
Mezzanine (hybrid of debt and equity)	Medium. Will bear some operational risks but not completion risks.	IRR, IRR target year	8–12%	Includes U.S. "tax equity"
Equity	High. Bears all project risks. Risk and return increase with debt "leverage"	IRR, Payback, other internal metrics	10–20%	Developer or private equity

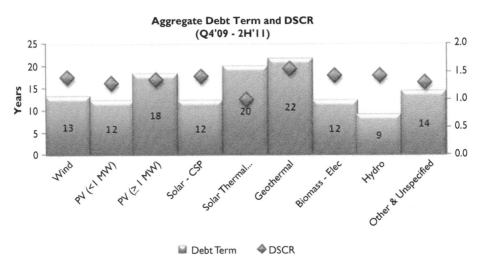

Figure 11.3 Debt term and DSCR as reported by NREL's Renewable Energy Finance Tracking Initiative (REFTI)[44] (Mendelsohn and Hubbell, 2012).

Second, investors in new renewable power generation projects are also able to accelerate the depreciation of a plant's capital investment via the five-year Modified Accelerated Cost Recovery System (MACRS) to defer federal taxable income. Figure 11.4 compares the percentage of the assets deductible annually under the five-year MACRS schedule, the five-year MACRS schedule with a 50% bonus

---

44 REFTI is a project designed to acquire financial data and project development insight form the renewable energy community. The project, run by NREL, polls the industry semi-annually on the cost of equity, debt terms, and other criteria and produces aggregated, confidential results which are presented via webinars and publicly available spreadsheets. REFTI data and presentations are available at: https:// financere.nrel.gov/finance/REFTI.

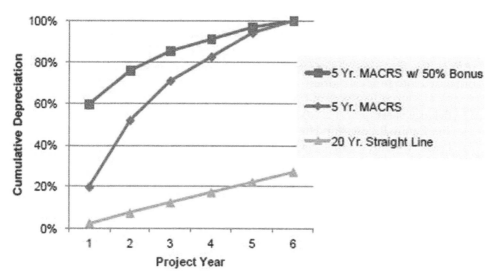

Figure 11.4 Speed of Depreciation under different tax schedules (Mendelsohn and Kreycik, 2012).

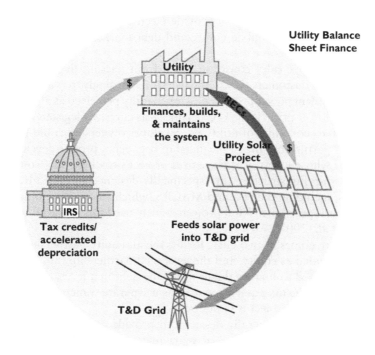

Figure 11.5 Utility balance sheet finance (Mendelsohn and Kreycik, 2012).

(renewable electricity plants generally are eligible in 2012), and a 20-year straight-line schedule.[45]

Together, the tax credits and the accelerated depreciation compose what is referred to as the "tax benefits" of a renewable project. To take advantage of these federal benefits, projects are structured in a variety of ways, which can generally be organized into four categories:

1) Balance Sheet
2) Tax Equity Partnership Flip
3) Lease Structures: Sale-Leaseback & Inverted Lease
4) Other

Balance sheet (also known as single owner): represents the direct investment by a developer in a project asset using the entity's general source of funds, or balance sheet. This source of funding is in contrast to "project finance," whereby investment in and cash flow out are specific to the project, and there is limited liability of the developer associated with payment of operating costs or debt service.

Utility-scale projects can be balance-sheet financed in one of two ways: utility-owned generation and developer balance sheet. Utilities – represented primarily by private investor-owned utilities (IOUs) – are generally stable, creditworthy entities and can attract capital at a favorable interest rate (even in tight credit markets). They also have a franchised authority to provide electric services and are experienced in raising capital in both the public equity and debt markets (Mendelsohn & Kreycik 2012).

However, complex rules constrain IOUs from passing the benefits of the ITC directly on to their customers (i.e., ratepayers). Accordingly, private developers, also known as independent power producers, are generally perceived as able to offer renewable energy at a lower price. If the ITC declines – as currently legislated – from 30% to 10% in 2017, the economics of utility vs. developer ownership could be altered.

Tax equity partnership flip: a partnership between a project developer and a tax equity investor whose investment monetizes project's tax benefits (Coughlin & Cory 2009). Partnership flip structures are specifically designed to take advantage of federal incentives, including the ITC and MACRS, which frequently cannot be fully utilized by the developer alone. Wind projects widely used this financial structure in the past (Harper et al., 2007).

The structure comes in two basic forms: the all-equity partnership flip, in which all capital is provided as equity, and the leveraged partnership flip, in which project-level debt is also used to finance the project. The debt in a leveraged partnership flip can be provided by the tax equity investor (as a separate tranche) or by a third entity depending on how the project is structured.

The flip structure is generally designed to provide the tax equity investor a pre-negotiated return in a set number of years (e.g., a 9% yield by year eight of the project). After that design goal is met, the annual stream of benefits (including tax

---

45 Only the first six years of the project life are shown in the Figure 11.4. All values are based on a "half-year" convention of the associated depreciation schedule.

benefits and cash) reallocates, or "flips," to the sponsor to reward the risk taken and work invested.

The following figure depicts an all-equity flip. Contributions and benefits in red are provided by /allocated to the developer; those in blue, to the tax equity investor. The "/" marks indicate a change in allocation benefits when a particular milestone or flip period is incurred. For example, as depicted in the figure, project cash may flow fully to the developer until the initial investment is recuperated, at which time all project cash is allocated to the investor until the pre-negotiated return is reached, at which time the allocation is split 95%/5% in favor of the developer until project termination. The other benefits, represented by accelerated depreciation and the ITC, can have unique allocations – open to negotiation among the parties.

Lease structures: Two basic lease structures are utilized to finance renewable power projects, the sale-leaseback and the inverted lease. In sale-leaseback financing, a project developer sells the project assets for cash and simultaneously signs a long-term lease with the investor. The developer then makes lease payments to the investor in exchange for the cash injection (Jacobs, 2009).

In an inverted lease (also referred to as a pass-through or master tenant lease), the developer leases the project to a tax equity investor and passes through the ITC to the tax equity investor. In turn, the tax equity investor (the lessee) sells the electricity to

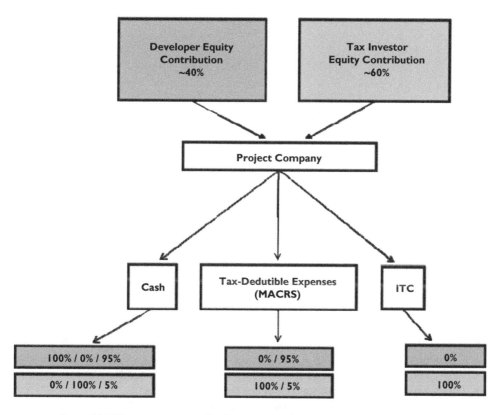

Figure 11.6 Tax equity partnership flip structure (Mendelsohn and Kreycik, 2012).

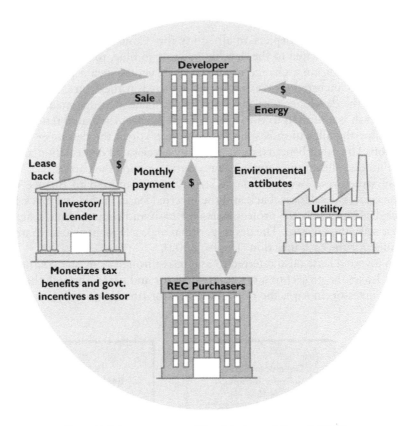

Figure 11.7 Lease structure (Mendelsohn and Kreycik, 2012).

the developer via a PPA arrangement (Tracy et al., 2011). The developer may operate the system on behalf of the investor pursuant to an operation and maintenance agreement.

Other: Projects may employ unique, one-off financing structures based on the risk appetites of the entities involved (including the developer, investor, and utility) as well as the investment climate at the time (e.g., the cost of debt and equity and perception of technology risk), including hybrid structures that combine attributes of the other structures referenced.

Third-party tax equity is generally provided by a very small set of large financial investors who:

- Have a substantial current and future tax appetite
- Have an internal team with financial acumen to engage in a complex project structure
- Are willing to hold their ownership interests in the projects for several years
- Are comfortable with an uncertain tax policy environment
- Are comfortable with illiquid investment classes (i.e., that tie up cash and cannot easily be resold) (Mendelsohn & Harper, 2012)

11.3 CAPITAL REQUIREMENTS

Asset finance is a global market, with U.S., European, and Asian projects all competing for capital. The total global need for asset finance including renewable energy and other infrastructure projects was $120 billion as of 2010, representing a five-fold increase from just six years prior (DBCCA, 2011).

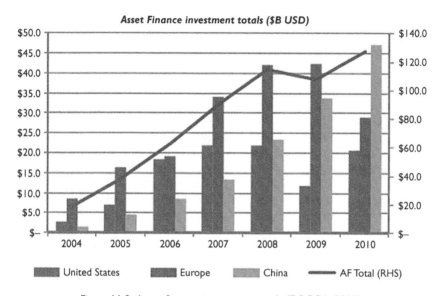

Figure 11.8 Asset finance investment totals (DBCCA, 2011).

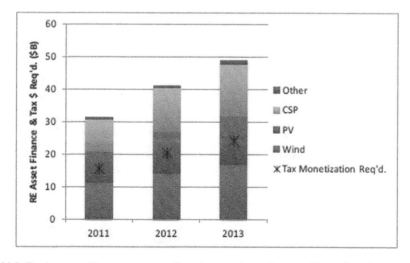

Figure 11.9 Total renewable energy asset financing sought and potential tax benefit monetization required (Mintz Levin/GTM, NREL) (Assuming 50% of Project Financing Needs).

The need for capital to finance U.S. renewable energy projects is expected to increase significantly over the next several years. A recent report forecasted that renewable energy project financial capital requirements will increase from $31 billion in 2011 to $49 billion in 2013 (Mintz Levin, 2012). Assuming 50% of the total capital required is represented by internal or third-party tax equity (i.e., to monetize tax benefits), that source of capital will increase from current market capacity of $15.5 billion to nearly $25 million in that two-year time frame assessed.

11.4   SOURCES OF CAPITAL

Global capital is provided from a wide range of sources, including various asset classes designed to support retirement (including pension funds and mutual funds held in retirement accounts such as 401K and Roth 401Ks). Figure 11.10 indicates the global assets held under various management categories.

In the U.S., however, renewable energy projects have had a very limited source of funds, primarily due to the complexities associated with monetizing tax benefits. Current rules dictate direct and active ownership, which require significant "due diligence" analysis by professionals with legal, engineering, electric transmission and other expertise, representing an expensive and time-consuming process. Accordingly, investment by pension funds, mutual funds, and private wealth has been very limited as there are very few investment vehicles – other than direct project ownership – that allow these types investors to take part in the industry (Mendelsohn & Harper, 2012). Furthermore, in the wake of the financial crisis which began in 2008 and

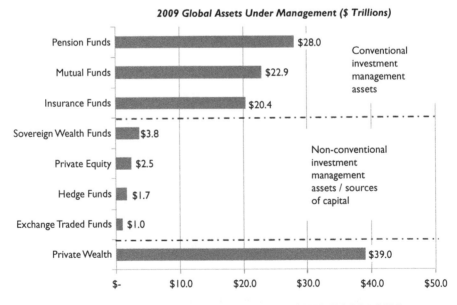

Figure 11.10 Global assets under management as of 2009 (DBCCA, 2011).

continues today, stringent lending rules are expected to constrain bank financing – widely utilized in Europe and Asia – for long-lived projects such as renewable energy assets.

11.5 DEVELOPMENT RISKS

Finance is a relationship of risk perception and reward expectation. A project with incremental risks will be required to provide greater return on the associated investment. The following table indicates a wide array of potential risks that are likely to be evaluated by lenders or investors to a given project.

These risks can represent an additional increase in the cost of capital, although any direct correlation would be impossible to draw. Roughly, the combination of perceived risks can be thought of as a stack that correlates to the final cost of debt, equity, or a weighted average cost of capital. The specific impact of a given risk on the cost of the financial components (debt, mezzanine finance, or equity) is subject to the nature of the financial structure applied.

Table 11.2 Risk and financial impact (Mendelsohn, 2011).

Risk	Explanation	Financial cost impact
Technology Risk	Any operational experience?	High
Developer Risk	Experience with technology, project size & type? Balance sheet strength?	Medium – perceived risk can lead to high reserve or guarantee requirements
Revenue Certainty	Is buyer credit-worthy? Are prices firm?	Medium – high
Duration of Revenue Support	Debt duration (tenor) no longer than PPA minus 2 years, shorter based on risk perception	Very high
Cost Certainty	Are components already purchased? Or are they subject to significant re-pricing?	Low
Currency Risk	Are costs/revenues in desired currency? Is currency stable against the dollar, euro, or yuan?	Low – can be hedged with currency swaps
Insurability	Is contractor/operator properly insured? What if something happens during construction	Medium
Site control	Does developer own the development location?	High in U.S.
Resource Certainty	Are long-term, quality measurements available? Variability in historic data?	"P" rating will impact debt service coverage ratios
Transmission Access	Interconnection & integration? Is intermittency an issue for grid stability?	Medium – High. Prior operational experience will be assessed
Political Risk	Are national/local governments stable; court action possible?	High

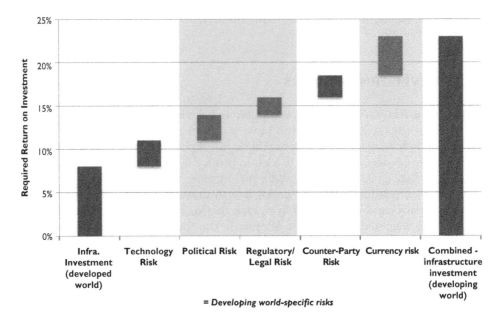

Figure 11.11 Required Return vs. Risk (DBCCA, 2011).

11.5.1   Risk mitigation techniques and policies

To mitigate risks and enable financing at a lower cost, numerous market responses and policies are available and roughly outlined in the following table. Market techniques range from long-term contracts (necessary, in particular, to support debt lending to the project) to advanced financial structures and securitization techniques such as asset-backed securities and real estate investment trusts.

Public support structures include tax credits and loan guarantees utilized by the U.S. for domestic deployment to foreign exchange risk mitigation, political risk insurance and similar credit enhancements designed for international investment (techniques applied by the Export-Import [Ex-Im] Bank and the Overseas Private Investment Corp).

Public support structures have various impacts on financing renewable energy projects, including the ability to lower the "installed cost" of the equipment, increase or decrease the associated transaction costs of acquiring capital, or alter the risk assumed by the market. In turn, public sector investment places taxpayer dollars at risk that the "investment" – via direct cash outlays or reduced tax revenue – will result in viable renewable energy manufacturing or energy production capacity.

Table 11.4 summarizes aspects of several common support structures. U.S. experience with loan guarantees and cash grants is discussed in Section 11.6.1.

Credit enhancements are a variety of investment support structures generally designed to leverage modest public investment to garner significant private (market)

Table 11.3 Private and public measures to mitigate risk and enable project financing (Mendelsohn, 2011).

	Category	Examples
Private (Market)	Long-term contracting	PPAs, leases
	Insurance	Construction, operation, debt obligation
	Advanced financial structures	Partnership flips, sale leasebacks
	Securitization	N/A (none currently for RE industry)
Public (Government)	Investment support structures (not exclusive – can combine)	Tax credits (ITC, PTC, MACRS) Loans or loan guarantees Grants/rebates Foreign exchange risk mitigation First loss reserves/credit enhancements Securitization underwriting (e.g., Federal National Mortgage Ass. for home mortgages)
	Market support structures (* exclusive – no need to combine)	Demand targets (RPS, Clean Energy Std.)* Guaranteed prices and purchases (Feed In Tariff)* Guaranteed markets (e.g. military contract) Emission targets* Net metering policies
	Indirect support structures (not exclusive – can combine)	Transmission build-out Labor education Infrastructure development (roads, shipping)

Table 11.4 Qualitative evaluation of different public support structures (Mendelsohn, 2011b).

Policy	Financing impact	Transaction costs	LCOE impact	Public risk taken
Tax Credits	Good	High	Moderate	PTC: low; ITC: moderate
Loan Guarantees	Excellent	Very high	High	Mfg.: High; Power Prod.: Low
Cash Grants	Good-excellent	Low	High	Moderate
Feed-In Tariffs	Good-excellent	Low	High	Low

investment. Figure 11.12 is a simple graphical representation of several public credit enhancement strategies ranging from a type of insurance product among a pool of projects referred to as "first loss" reserve, to co-investment equal in risk allocation as private equity or as a mezzanine product which serves to protect the lenders in order to attract that type of capital.

11.6   POLICY IMPACT ON ASSET FINANCING

Public policy is critical to renewable energy development, and the ability to raise capital for renewable energy projects. For example, Germany is widely recognized as

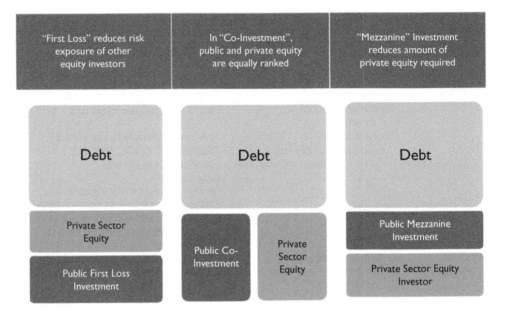

Figure 11.12 Credit enhancement structures (DBCCA, 2011).

a world leader in solar energy deployment, primarily due to its combination of policies that support deployment. The country installed 7.5 gigawatts (GW) in 2011, with roughly 3 GW added in December alone (SolarServer, 2012). That represents roughly 28% of total 2011 worldwide installations of 28 GW – impressive for a country that represents just 3% of global electricity sales and slightly more than 1% of global population (Mendelsohn, 2012a).

The country has effectively pushed down the price of its solar installations through stable purchasing and financing policies. See Figure 11.13, which depicts the downward feed-in-tariff (FIT) pricing trend along with the consistent procurement of total yearly solar power. As of March 2012, the solar FIT in Germany ranges from 17.9 eurocents/kWh for ground mount installations to 24.4 eurocents/kWh for rooftop systems up to 30 kW (equivalent to $0.23/kWh – $0.31/kWh at current conversion rates).

Germany also directly supports the financing of solar installations. The federally owned KfW bank lends to individuals and community groups to facilitate renewable energy and energy efficiency investments. In 2010, KfW financed 40% of photovoltaic installations. The bank recently committed to lending 100bn ($140 billion) over the next five years. Interestingly, KfW was listed as #1 among the world's 50 safest banks in 2010 by Global Finance magazine (Global Finance, 2010).

Germany also offers guaranteed access to the grid for solar installations. When combined, the FIT, lending policy, and grid access create a low-cost environment to develop renewable energy. The response has been significant market investment

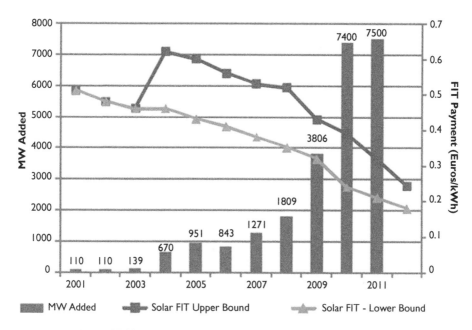

Figure 11.13 PV installations and FIT in Germany (DBCCA, NREL).

Table 11.5 Required FIT payments assuming German installed cost and cost of capital (Mendelsohn, 2012).

Location	Capacity factor (1 axis tracking)	Calculated solar FIT
Frankfurt, Germany	12%	$0.230
U.S. Locations		
Los Angeles, CA	23%	$0.123
Richmond, VA	20%	$0.141
Honolulu, HI	25%	$0.111
International Locations		
Sao Paulo, Brazil	17%	$0.173
Manila, Philippines	15%	$0.178
Cairo, Egypt	24%	$0.128

Capacity factors calculated via NREL's System Advisor Model (SAM).

and a large industry of module and component manufacturers, installers, and related services.

In contrast, the U.S. has installed far less solar electric capacity, even with better solar resource and far higher total electrical requirements. Arguably, American policies have not yet driven down the cost of solar energy to the prices available in

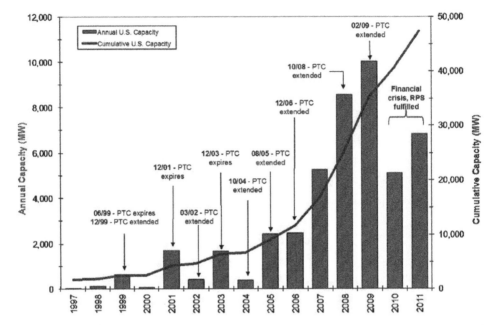

Figure 11.14 U.S. Tax credit policy impact on wind project development. (*Source:* American wind energy association, "Industry statistics" and schwabe, P.; James, T.: Cony, K. "U.S. renewable energy project financing: Impact of the global financial crisis and federal policy," international sustainable energy review).

Table 11.6 Value of current and future tax benefits (Mendelsohn, 2012b).

Year	0	1	2	3	4	5	6
Project Cost	$100						
Tax Rate	$35						
IRR Target	10%						
Current Tax Equity Requirement							
ITC	30%						
ITC Value	$30						
Depreciable Basis	$85						
5-Yr. MACRS + Bonus Schedule	60.0%	16.0%	9.6%	5.8%	5.8%	2.9%	
Depreciation Value (schedule × basis × tax rate)	$18	$5	$3	$2	$2	$1	
Total Tax Benefit (depreciation value + ITC)	$48	$5	$3	$2	$2	$1	
Tax Equity Inv. that Earns 10% IRR on Tax Benefits ($52)							
Future Tax Equity Requirement							
ITC	10%						
ITC Value	$10						
Depreciable Basis	$95						
5-Yr. MACRS (no bonus)	20.0%	32.0%	19.2	11.5%	11.5%	5.8%	
Depreciation Value (schedule × basis × tax rate)	$7	$11	$6	$4	$4	$2	
Total Tax Benefit (depreciation value + ITC)	$17	$11	$6	$4	$4	$2	
Tax Equity Inv. that Earns 10% IRR on Tax Benefits ($35)							

Germany, although the Department of Energy's (DOE) Sunshot Initiative is aimed at bringing the cost down dramatically over the next several years.[46]

By applying solar resource data from Frankfurt, Germany, we can estimate the installed costs of a large-scale solar system to be $1.83/watt based on Germany's utility-scale FIT value.[47] Table 11.5 analyzes solar FITs in alternative locations (with superior solar resources), assuming the low installed cost available in Germany. No other tax credits or depreciation benefits were assumed.

As indicated in the analysis, various U.S. and international cities offer superior solar resource to Frankfurt, Germany and thus low FIT values, if installed costs can be driven down to levels implied by Germany's current FIT value. For example, in the U.S., FITs from 11 and 14 cents/kWh could support large-scale PV installations from Hawaii to Virginia, respectively.

11.6.1   U.S. policy to induce renewable energy finance

In the U.S., two primary policy inducements are used to support renewable energy deployment: tax credits and renewable portfolio standards.

Tax credits

In the U.S., policy consistency is a primary concern among the renewable energy development and investment communities. The tax credit is frequently cited as an inconsistent policy, although it is also widely recognized as critical to a project's success and ability to raise capital (Mendelsohn & Harper, 2012). For example, the PTC has expired, or been renewed in the last quarter-year of its existence, at least seven times over the last 15 years. Figure 11.14 displays wind capacity development and legislative support of the PTC.

The investment and production tax credits combined with accelerated depreciation benefits are sometimes referred to together as "tax benefits". Over the next several years, the components of the tax benefits decline, but still represent a sizable component of the total project cost. As referenced above, the ITC will decline from 30% to 10% of eligible capital costs in 2017. In addition, at the end of 2012, the current 50% "bonus" depreciation expires. Accordingly, the need for tax equity will ease, but may still be a difficult hurdle for many developers.

Table 11.6 assesses how much tax equity will be needed after these adjustments occur based on the value of tax benefits under the current and future incentive structures for a $100 solar project.[48] The value of the tax benefits equals $52, or 52% of the initial project costs under the current incentive levels. In 2017, after the ITC and depreciation benefits revert to their former levels, the value of the tax benefits will drop to 35% of initial project costs.

---

46 DOE has a goal of $1/watt installed for utility-scale systems by 2017. See http://www1.eere.energy.gov/solar/sunshot/index.html

47 Assuming a 8% weighted average cost of capital, seven year straight-line depreciation, and 1-axis tracking. Low-cost funds available from Germany's KfW bank were excluded as that support structure isn't available in the U.S.

48 Assuming a 35% tax rate and a 10% internal rate of return.

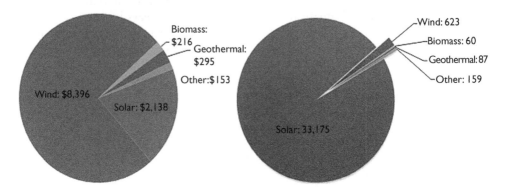

Figure 11.15 §1603 Program total funding ($ millions) and §1603 Program total projects awarded (Adapted from the U.S. Department of the Treasury).

Section 1603 cash grants

Pursuant to Section 1603 of the American Recovery and Reinvestment Act, the federal government allowed taxpayers to elect to receive a cash payment in lieu of the PTC or ITC (i.e., a grant). The 1603 program has had a dramatic impact upon renewable power development. Through March 29, 2012, the §1603 Program had awarded $11.2 billion to over 34,000 projects.[49]

Assuming §1603 awards equal 30% of total project costs, the program has supported, up through March 2012, some $37.3 billion in total investment. A projected 16.5 gigawatts in capacity, expected to produce 42,000 gigawatt-hours (GWh) annually, have been attributed to the §1603 Program.[50] A wide range of technologies have received cash awards under the program, including wind, solar, hydroelectric, geothermal, biomass, and fuel cell technologies.[51]

According to a recently-completed analysis, the 1603 program provided a wide-range of benefits over the ITC program it replaced (Mendelsohn & Harper, 2012), including:

- Increased speed and flexibility
- Lowered transaction and financing costs
- Stretched supply of traditional tax equity
- Supporting smaller developers (including new entrants) and innovative technologies that were less capable to tap tax equity markets
- Improved economics of most renewable energy projects

49 Overview and Status Update of the §1603 Program. U.S. Treasury. March 29, 2012. http://www.treasury.gov/initiatives/recovery/Documents/Status%20overview.pdf. Accessed April 23, 2012.
50 U.S. Treasury. 2012, *op cit.*
51 Awardees represented in "List of Awards" spreadsheet available on Department of Treasury 1603 program website. March 13, 2012. http://www.treasury.gov/initiatives/recovery/Pages/1603.aspx Accessed April 24, 2012. The spreadsheet does not indicate all projects – some are grouped by awardee.

- Allowing use of more debt and lowering developer or project cost of capital
- Generally supporting an extensive build-out of renewable power generation projects

The analysis also found three potential and not mutually exclusive outcomes expected as the §1603 Program ends:

a) Smaller or less-established renewable power developers, especially those with smaller projects, are expected to have more difficulty attracting needed financial capital and completing their projects. This is likely to lead to industry consolidation as well-funded developers acquire smaller firms.
b) Development of projects relying on newer or more "innovative" technologies that have little operational track records will likely slow, as many tax equity investors are seen as highly averse to technology risk in the projects they fund.
c) Projects relying on tax equity are usually more expensive to develop due to the transaction costs and potentially higher yields required to attract tax equity. In the short term, such cost increases may slow new investment in renewable power projects by reducing developer returns. Over time, due to absent new tax equity investors entering the market or offsetting reductions in capital cost, the incremental cost to project financing might increase the cost to utilities and end-users who procure renewable power. (Ibid)

Loan guarantees

The DOE loan guarantee program made a total of $16.16 billion in loans under section 1705, 88% of which (or $14.23 billion) were designed support power generation projects. Loans supporting power generation projects represent a very different risk profile than those supporting manufacturing facilities (such as Solyndra, which failed). According to Ken Hansen, an attorney with Chadbourne Parke and the former General Counsel of the Ex Im Bank, the risk of default for loans supporting generation projects is extremely low given that they (i) hold long-term contracts with credit-worthy utilities, (ii) often have production guarantees from the equipment manufacturer, and (iii) often have construction-cost guarantees from international construction firms. In the case of the projects that embody some form of innovative technology, the functionality of that technology is thoroughly reviewed by DOE and its independent engineers before DOE agrees to finance the project (Mendelsohn, 2011c).

DOE required a credit subsidy cost of 11.7% or $1.88 billion, which was funded out of a pool of money established under ARRA, originally set at $6 billion, but later reduced to $2.44 billion. As much as 22% of the total $2.44 billion credit subsidy available may already have been consumed, assuming the government does not get money back from the liquidation of assets of failed projects.

On a very simplistic basis, the government credit subsidy should lead to roughly 12 terawatt per hour (TWh) of energy per year according to DOE estimates, or 212.5 TWh over the next 20 years, assuming 11.7% of the capacity is lost to bankruptcy and ignoring any productivity declines. If all of the credit-subsidy put aside for those loans is consumed, taxpayers will pay $7.84 for every megawatt per hour

Table 11.7 Credit subsidy per MWh of production (Mendelsohn, 2011c).

Program	LG amount ($B)	Assumed credit subsidy ($B)	MW (ac)	Annual generation per DOE (GWh)	Assumed 20 years of generation less 11.7% (GWh)	Calculated credit Subsidy per MWh over 20 years
Photovoltaic Electric Generation	6.14	0.72	2,272	4,730	83,528	$8.60
CSP Electric Generation	5.86	0.69	1,243	3,623	63,982	$10.72
Solar Manufacturing	1.28	0.15			0	
Wind Generation	1.70	0.20	1,025	2,188	38,640	$5.11
Other Technologies*	1.19	0.14	180	1,492	26,349	$5.27
Total	16.16	1.88				
Electric-generating only	14.23	1.67	4,720	12,033	212,499	$7.84

* Other Technologies includes a mix of electric and non-electric generation.

(MWh) of expected electricity production over the next 20 years, not including any discounting.

In contrast, energy cost savings from the program may far exceed the expected costs. In fact, low-cost debt available under the loan guarantee program can reduce the levelized cost of the generated electricity about $20/MWh for photovoltaic projects and $29–$37/MWh for projects using Concentrating Solar Power (CSP) technologies (Mendelsohn et al., 2012).

11.7   PATH FORWARD: SECURITIZATION?

One potential path to improving the availability and cost of capital to deploy renewable generation projects is securitization. Securitization, in a general sense, transforms illiquid financial assets into tradable investment products in order to attract capital from a wide array of sources (Mendelsohn, 2012b). Similar to the way a mutual fund pools stocks, bonds or other assets, securitization mechanisms such as master limited partnerships (MLPs), real estate investment trusts (REITs), and asset-backed securities (ABS) offer the opportunity to enable investment through more liquid and transparent investment vehicles (Mendelsohn, 2012c).

However, significant regulatory and market barriers remain to the application of these mechanisms to renewable energy project finance. For example, MLPs are generally considered not applicable to solar and wind project development (the MLP-enabling legislation refers to deplete-able natural resources such as natural gas or oil transportation projects). REITs may be available to solar property if the property is bundled with real assets or if solar properties (or some portion thereof) are specifically ruled on as real property (Feldman, et al., 2012). Additional risks – such as long-term production, transmission access rights, or customer default – also complicate the attractiveness of the potential investment, and could remain significant barriers to securitization success.

Chapter 12

Business: Insurance and risk management

Scott Reynolds

A critical component of project success is the management of risk.

For power generation projects, risk assessment starts with an understanding your business, risk profile and risk appetite. The risk assessment process evaluates both the likelihood of the unwanted event as well as the consequence. The process is often qualitative in nature – i.e. relative risk ranking, however, a quantitative risk assessment can be undertaken if needed. It is recommended the risk assessment be completed before capital has already been committed as this allows for improved financing opportunities as well as the optimization of costs over the life of the asset. Additionally by addressing risk and insurance issues right from the start, contract negotiations can be targeted, lender/finance parties can be satisfied and the most cost effective risk solutions can be more readily identified and implemented.

12.1 RISK ANALYSIS

Risk management is a system for planning, organizing and controlling the resources and activities of an organization in order to protect itself from the adverse effects of accidental loss. Although there are different methodologies for risk management, the core components of any risk analysis is made up of the following:

1) Identify loss exposures
2) Analyze loss exposures
3) Examine feasibility of risk management techniques
4) Select the appropriate risk management techniques
5) Implement selected risk management techniques
6) Monitor results and revise the risk management program

A simple tool designed to assist in consistent evaluation of risk to a defined criteria is a Risk Matrix. The objective is to reduce identified high and significant risks to more acceptable levels of moderate to low risk. Risk improvement action can be prioritized to maximize corporate risk management objectives within a necessarily constrained fiscal budget.

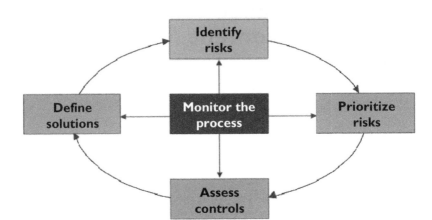

Figure 12.1 Risk analysis flowchart.

The analysis to be performed should include a full understanding all of a projects risk exposures. This includes:

- Analyze Financials (Annual Statement or Balance Sheet)
- Assess and analyze contractual risk allocation. Highlight unallocated risk.
    - What are the terms of the Credit Agreements, O&M contracts and/or EPC contracts?
    - What are the required limits, sub limits, deductibles and terms in each of those contracts?
    - Are there specific loss payable endorsements required for individual projects?
    - Ensure consistency of risk and insurance clauses in major contracts
- Assess and analyze insurable risk, design insurance program, highlight uninsurable risk
    - Stratify the Project Cost to determine the exposure to loss
    - Risk Retention
        - What are project economics and tolerances for retaining risk?
    - Transmission & Distribution Lines, Transformers and Inverters
        - What are the terms of the Interconnect, and do we need to adjust the insurance protection by project with specific endorsements?
    - Natural Catastrophe Perils
        - Determine the Probable Maximum Loss Evaluation and Maximum Foreseeable Loss Evaluation
            - Establishing realistic Property Damage, Business Interruption, and Extra Expense Loss Estimates to help determine adequate coverage limits.
                - Named Windstorm (Hurricanes)
                - Flood
                - Earthquake

Table 12.1 Probability and consequence risk analysis matrix.

	Consequence				
Probability	Insignificant	Minor	Moderate	Severe	Catastrophic
Very Frequent	High	High	Urgent	Urgent	Urgent
Frequent	Moderate	Hight	Hight	Urgent	Urgent
Moderate	Low	Moderate	Hight	Urgent	Urgent
Seldom	Low	Low	Moderate	Hight	Urgent
Rare	Low	Low	Moderate	Hight	Hight

- Do we need to adjust the flood, earthquake, wind coverage to comply with Lender requirements?
- Loss of Profit Concerns
  - Business Interruption and Period of Indemnity
    - Are 12 months sufficient to rebuild the project in the event of a major or catastrophic event? Is a longer period of Indemnity needed?
  - Extra/Expediting Expense
    - Is the sub-limit provided in the overall insurance policy sufficient for this project?
  - Contingent Business Interruption
    - What is the Project's exposure to loss if customers and/or suppliers are unable to perform?
      - Insurance Underwriters will typically extend coverage to Named suppliers and customers only
  - Ingress Egress
    - What is the Project's exposure to loss from the inability to enter or exit the Project location?
    - Are the terms provided by the insurance consistent with the exposure for the project?
- The Individual Plant
  - Technology, equipment experience, operating hours and loss experience
- Review the Historical Loss Record
  - Kind/type of losses
  - Is there a frequency or severity issue
  - Impact varying deductible levels would have on the loss record
- Examine available engineering/loss control reports
  - Review outstanding recommendations
- Determining the Projects Risk Appetite
  - Retention vs. Risk Transfer
  - Price vs. Coverage
- Has the accumulation of risk, specifically catastrophic and warranty risk, been measured, understood and addressed?
- Standard Insurance Exclusions
  - Do they need to be adjusted for specific projects?

12.2 PROJECT RISK

For power generation projects there are typically three main types of risk exposures:

1) Manufacturing, managing design, testing, sub-assembly manufacture, and quality control.
2) Installation/Construction at the project sites, managing subcontractors, and start-up responsibilities.
3) Operation & Maintenance of the power project, retaining efficacy risk and responsibility for plant maintenance and security.

Table 12.2 provides a good overview of the risks of a Power Generation project.

12.2.1 Risk mitigation

Once the Risk Analysis is finished, one can identify the necessary countermeasures to mitigate the calculated risks, and carry out cost/benefit analysis for these countermeasures.

There are four steps pertaining to each of the identified risks:

1) Mitigating the risk by implementing the recommended countermeasure
2) Accepting the risk
3) Avoiding the risk
4) Transferring the risk by purchasing insurance

The objective is to reduce identified high and significant risks to more acceptable levels of moderate to low risk.

12.2.2 Anticipated insurance coverage's

Table 12.3 provides an overview of the anticipated Project Insurance coverage's.

Achieving financial close

The key to achieving financial close and notice to proceed is to ensure appropriate provision of protection and risk management. The lender/finance parties will need to be satisfied that risks have been adequately identified, quantified, eliminated, reduced or transferred.

As your portfolio grows

As your project portfolio grows, a Master Insurance Program may be appropriate to establish, covering all of your projects under a single risk management and insurance program. The advantages of a Master Insurance Program include a predictable underwriting process, pre-agreed premium rates along lower costs and broader coverage

Table 12.2 Risk overview for power generation projects.

Underwriting considerations	Manufacturing	Installation	Operation & maintenance
Developer, OEM, Installer and O&M	Business Interruption	Loss record and experience of contractors	Experience level will affect availability and price of insurance products
	Specialty Tool & Die		
	Ordinary Payroll	Builders Risk/CAR (aka Installation Floater)	
	Technology		Either "Total Insured Value" (TIV) or "Loss Limit" approach
	Loss control for Workers Compensation	Delayed Start-Up (DSU)	
		• Gross Earnings	Separate declarations, limits and sub-limits, loss payees, mortgagees
	Fire Loss Prevention	• Additional expenses, fixed operations and maintenance expenses	
		• Advance Loss of Profits	Loss of Profit: Business Interruption, Contingent Business Interruption
		• Liquidated damages	
			• Match to lender and PPA requirements
		Transit coverage, including DSU	• Coverage for Production Tax Credits and other renewable credits
		Explanation of testing and acceptance procedures	
			• Replacement lead time for key pieces of equipment/spares management
		Consistency in subcontractor coverage	
			Calculation of Maximum Foreseeable Loss (MFL) and Probable Maximum Loss (PML)
			Transmission and distribution lines
			Substations and transformers
			Legal (Regulatory Requirements)
			Safety (Personal and Process)
			Liability (Internal and Third Party)
			Fleet Exposures Environmental Goodwill/Public Relations (Reputational Risk)

Table 12.3 Anticipated project insurance coverages.

Milestone for policy Inception	Insurance coverage	Purpose	General level of coverage	Cost basis	Who provides the policy
Project Development	General Liability	Protects from losses arising out of your legal liability to third parties	$1m per occurrence	Construction: Project Cost Operational: MW Output of Facility	The Project and All Contractors. Contractors should be primary especially during construction
	Automobile Liability	Protects from losses arising from owning or operating an automobile	$1m per occurrence	Cost per vehicle or hire	The Project and All Contractors.
	Umbrella Liability	Provides higher limits of liability for underlying coverage (GL, Auto, Employers Liability)	Minimum of $10 m with a typical maximum of $50 m	Based on underlying premium	The Project and All Contractors
	Pollution	Provides coverage for environmental claims arising from your operations on site	$5 m per occurrence (multi-year policies available)	Site conditions and information available for underwriting	The land owners or lessee. Contractor may be required to purchase additional coverage
	Executive Risk (D&O, Employment Practices Liability, etc.)	Protects individuals and their personal assets as fiduciaries of the Project and provides defense costs	Typically $1 m to $5 m depending on asset size & investment structure of company	Asset size and employee count	The Project
Financial Close or Commencement of Construction	Builders Risk & Delay in Start Up	Protects the Owner, Lender, and Contractors from first party physical damage and loss of revenue claims as a result of a covered loss	Replacement cost of project & min of 1 year of lost profits and continuing expenses	Based on the values as reported for the entire duration of the construction	The Project

	Coverage	Description	Limit	Basis	Covered
	Marine Cargo Delay in Start Up	Coverage for on and off-shore property. Protects the Owner, Lender, and Contractors from first party physical damage and loss of revenue claims as a result of a covered loss	Replacement cost of largest values shipped at any one time plus min. 1 year of lost profits and continuing expenses	Value of shipments & loss of profits	The Project – both Marine and loss of profits (DSU) The Manufacturer – typically is responsible for the primary Marine coverage
	Professional Liability	Protects The Project from a wrongful act committed by the Designer, Engineer, or Construction Manager from their work. Does not apply to property damage	Typically a range of $1 m to $5 m depending on the scope of work	Based on the firm's own exposure and experience	Designer, Engineers, Construction manager
Hire Employee	Workers Compensation and Employers Liability	Workers Comp is usually the sole remedy for injured employees. Coverage to include USL&H and Jones Act as necessary	Statutory for Workers Compensation and $1 m for Employers Liability	Based on payroll and type of employee	The Project (to the extent it has employees) and All Contractors
Commercial Operations	Property & Business Interruption	Protects the Owner, Lender, and Contractors from first party physical damage and loss of revenue claims as a result of a covered loss. Coverage for on and off-shore property	Replacement cost of the project plus a min of 1 year of lost profits and continuing expenses	Based on the values as reported	The Project

through economies of scale. The predictable underwriting process can then be used in your planning process and financial models.

12.3   CONCLUSION

In summary, understanding the risk profile of a project or series of projects is critical to the success of a project. Ensuring the identification and management of the risk enables you to evaluate the likelihood of an unwanted event as well as the potential consequence. The objective is to reduce identified high and significant risks to more acceptable levels of moderate to low risk. A simple tool designed to assist in consistent evaluation of risk to a defined criteria is the Risk Matrix. Finally, the assessment of risk for power generation projects needs to be addressed right from the start as this allows contract negotiations to be more targeted, lender/finance parties to be satisfied that the project risks are being properly addressed and cost effective risk solutions can be more readily identified and implemented. By challenging every aspect of project risk, the probability of success is greatly enhanced.

Chapter 13

Renewable energy and Mexico – an emerging market with promise

Pilar Rodríguez-Ibáñez

While California, Spain, and even New Jersey saw widespread and rapid installation of solar energy systems in a short period of time, this was largely motivated by policy change instead of pure economic gain. Conversely, Mexico is among a growing list of countries where the cost of solar energy produced may be lower than the cost of energy available from their current base of generation. While the long term benefits of renewable energy are often overlooked by the opportunistic market. Countries in this situation may experience an organic boom in the number of solar power systems installed due to these pure costs to cost energy price opportunities. It is important in this dynamic global market to be diligent to new markets and opportunities. Hence, in this chapter we present an analysis of the Mexican energy market including its history and its high potential due to its solar resource and proximity to the U.S. market. It is just one of several countries in the world that may be at a tipping point. Any could be "the next big market". Further, the manner of study shown here for this market is relevant for any market investigation and we suggest the reader conduct similar studies periodically to discover other diamonds in the rough.

13.1 THE LAW ON THE USE OF RENEWABLE ENERGY AND ENERGY TRANSITION FUNDING (LAERFTE)

Although the topic of renewable energy in Mexico is not new, as the founding of the Mexican National Solar Energy Association (ANES),[52] occurred more than 30 years ago, it was not until the 28th of November 2008 that the Law on the Use of Renewable Energy and Energy Transition Funding (LAERFTE) was put into effect by the Official Journal of the Federation, which for the first time granted renewable energy formal recognition under Mexican law.

The law aims to "regulate the use of renewable energy[53] sources and clean technologies to generate electricity for purposes other than the public power and the creation

---

52 Asociación Nacional de Energía Solar. [Online] Available from: http://www.anes.org/anes/ [Accessed April 2012].

53 The LAERFTE defines a renewable energy as those whose source lies in the natural phenomena, processes or materials likely to be transformed into usable energy for humanity, which regenerate naturally, by what is available on a continuous or periodic basis. It lists them: a) the wind; (b) solar radiation; (c) the movement of water in natural or artificial channels; (d) the ocean energy in its various forms;

of national strategy and instruments to finance the energy transition" (Art. 1). To ensure the use of renewable energy, the Federal Government, through the Ministry of Energy, will develop and coordinate the implementation of the Special Program for the Development of Renewable Energy. This program aims to promote social participation in the planning, implementation and evaluation of the program, as well as to set specific goals for the use of renewable energy and strategies to achieve them (Art. 11).

The LAERFTE indicates that the Energy Regulatory Commission (CRE) shall have the responsibility to issue regulations, guidelines, and methodologies governing the generation of electricity by renewable and sustainable means. The Commission will establish the regulatory instruments for the calculation of the consideration for the services provided by suppliers and generators[54].

In order to promote the development of renewable energy, the Secretary of Energy may enter into agreements and coordination arrangements with the governments of the Federal District, States, and, when appropriate, the municipalities. It should be noted that the law stipulates that projects generating electricity from renewable energy with a capacity greater than 2.5 megawatts (MW) should ensure the participation of local and regional communities through meetings and public consultations (Articles 8 and 21).

In another key area, the law provides that the Federal Executive Committee shall annually present the National Strategy for Energy Transition and Sustainable Use of Energy. This strategy is aimed primarily at promoting the use, development and investment of renewable energy to reduce Mexico's dependence on oil as a primary energy source. The Strategy has a horizon of 15 years, with a Fund for Energy Transition and Sustainable Energy Usage to be used to financially support projects that meet its objective (Article 27).

Finally, the LAERFTE requires the Secretary of Energy to report on progress in the transition from traditional to renewable energy as well as sustainable energy usage goals. As of June 1, 2011, two articles were added to transitional law which provides that the Secretary of Energy set a target maximum share of electricity generation by fossil fuels at 65% to be achieved by 2024, 60% in 2035 and 50% in 2050.

In addition to LAERFTE, the Mexican legal framework regulating renewable energy sources is supported by the following laws:

1) Public Service Law of Electricity (Articles 3 and 36)
2) The Interconnection Agreement for Renewable Energy Sources
3) The Law on the Promotion and Development of Bioenergy (Article 12)

(e) heat from the geothermal fields and (f) the bioenergy (Art. 3, II). Cámara de Diputados. LXI Legislatura. (n.d.) *Ley para el Aprovechamiento de Energías Renovables y el Financiamiento de la Transición Energética.* [Online] Available from: http://www.diputados.gob.mx/LeyesBiblio/ref/laerfte.htm [Accessed April 2012].

54 These attributions that the LAERFTE gives the CRE were published in the Official Journal of the Federation conventions and model contracts to regulate power generation according to the capacity, which are the following: 1. contract of interconnection for sources of renewable energy and small-scale cogeneration systems; 2. Contract of interconnection for sources of renewable energy and cogeneration in medium-scale systems; 3. Contract of interconnection for electric power generation with renewable energy or efficient cogeneration plants; and 4. Contract of interconnection for source of hydroelectric power.

Table 13.1 Participation targets for maximum fossil fuel electricity generation (Cámara de Diputados. LXI Legislatura., n.d. (a)).

Cumulative objective	Quantitative objectives: % of fossil generation targets
Reduce Mexico's dependence on hydrocarbons as the primary source of energy.	2024: Maximum Generation: 65%
	2035: Maximum Generation: 60%
	2050: Maximum Generation: 50%

4) The Law on Income Tax (Article 40)
5) The General Law of Ecological Equilibrium and Environmental Protection (Article 64)
6) The National Water Law (Articles 80 and 81). At the local level, the states of Chiapas, Guanajuato, Coahuila, Durango and Sonora have created laws to promote renewable energy[55].

The policy for the development of renewable energy

During the decade from 1994 to 2004, the policy implementation and motivation to increase renewable energy was marginal[56]. However, during the presidency of Vicente Fox (2001–2006), the country's energy policy began to be geared towards boosting the generation of electricity from renewable energy sources, mainly water, wind and geothermal. By the end of 2005, the Energy Regulatory Commission (CRE)[57] authorized permits to 54 individual projects in terms of supply, cogeneration and export. As displayed in Table 13.2, of the 54 projects authorized for permits, seven were permits for private projects of self-supply renewable energy from wind technology (Sandoval, Bosl, & Eckermann, 2006):

Today's renewable energy policy in Mexico is derived from the 2007–2012 National Development Plan (NDP), the Energy Sector Program 2007–2012, as well

---

55 The Congress of the State of Guanajuato created the law for *the Fomento del Aprovechamiento de las Fuentes Renovables de Energía y Sustentabilidad Energética para el Estado y los Municipios de Guanajuato*, published on 8 November of the year 2011. The Congress of the State of Coahuila created the law of *Ley de Fomento al Uso Racional de la Energía* in the year 2007, which consists of 18 articles. The law for the *Ley para el Fomento, Uso y Aprovechamiento de las Fuentes Renovables de Energía del Estado de Durango y sus municipios* was published in the official newspaper No. 1 dated 3 January of 2010. Finally, the *Ley de Fomento de Energías Renovables y Ahorro de Energía del Estado de Sonora* was published on August 27, 2009 and consists of 26 articles.
56 According to the Ministry of energy, the participation of the geothermal and wind energy grew hardly 1% during this period.
57 The CRE regulates the natural gas and electricity industries, grants the permits for the generation of energy, adopted the framework for the provision of energy contracts, and methodologies for the calculation of tariffs for private energy service providers. Cámara de Diputados. LXI Legislatura. (n.d.) *Ley de la Comisión Reguladora de Energía*. [Online] Available from: http://www.diputados.gob.mx/LeyesBiblio/ref/lcre.htm [Accessed April 2012].

Table 13.2 Permits issued for renewable energy generation
(Sandoval et al., 2006).

Technology	Permits	Capacity (MW)	Energy (GWh/yr)
Wind	7	956.73	3,645.31
Hydro	12	159.08	736.33
Cane Biogas	4	70.85	205.30
Biogas	3	19.28	120.80
Hybrid*	28	248.68	475.40
Total	54	1,454.62	5,183.14

* Renewable sources with fossil fuels.

as the LAERFTE, which provides for the establishment of the Special Program for the Development of Renewable Energy.

The Energy Sectoral Program proposed, as one of its objectives, to promote the use of renewable energy and biofuels. Its strategies are: to encourage private investment in the creation of companies dedicated to the design of components and equipment using renewable energy development schemes, leverage or promote funding for renewable energy sources, expand the coverage of electricity in remote communities using renewable energy, and support research and training of human resources in the various fields of renewable energy.

The government of President Felipe Calderon (2006–2012) has opted to promote hydro, wind and solar as renewable energy sources to generate electricity. This is supported by the Special Program for the Development of Renewable Energy, which set a goal that the percentage of installed capacity by renewable energy sources in the country to be increased from 3.3%, in of the national total 2008, to 7.6% in 2012.

According to data from the Energy Department, as of August 2010, Mexico had an installed capacity of renewable energy of 2.365 MW, equivalent to 4%, with a needed additional capacity of 3.6% to meet the target set for 2012. Of current installed capacity, 41% (965 MW) is geothermal, 21% (493 MW)[58] is wind, 19% (459 MW) is biomass, and 18% (416 MW) is water generated in power plants with capacity below 30 MW (see Figure 13.1).

Under the current plan for these sources of renewable energy, wind power would take a substantial leap in capacity. It is planned to rise from 0.15% in 2008 to 4.34% for 2012. However, solar has lagged far behind wind, with the installed capacity reaching an almost marginal 28.62 MW as of 2010.[59] In the next section we will discuss in more detail the progress in the introduction of these two types of energy to the Mexican market.

58 Global Wind Energy Council. (n.d.) Regions. Mexico. [Online] Available from: http://www.gwec.net/index.php?id=19 [Accessed April 2012].
59 Asociación Nacional de Energía Solar. (2010) Balance de Energía. [Online] Available from: http://www.anes.org/anes/index.php?option=com_wrapper&Itemid=13 [Accessed April 2012].

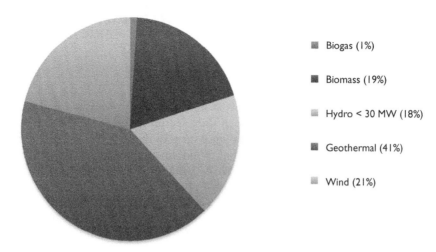

Figure 13.1 Total installed capacity by renewable energy in Mexico. (SENER-Mexico).

In addition to the federal laws discussed, there are other programs that are part of Mexican politics regarding renewable energy that have been promoted by the current federal administration. The main programs are: The Program for the Promotion of Solar Water Heaters (PROCALSOL)[60] and the Green Mortgage Program (INFONAVIT)[61].

Its momentum in the residential, commercial, industrial and agricultural business have caused PROCALSOL to estimate install 1.8 million square meters of solar heaters by the end of 2012 in Mexico[62]. The Green Mortgage is another monetary incentive awarded by the federal government program INFONAVIT to beneficiaries who purchase a home that meets federal criteria regarding the implementation of efficient technologies that reduce water consumption, electricity and natural gas.

Finally, the Mexican government offers economic stimulus to those who purchase renewable energy related instruments and machinery. Article 40 of the Law on Income Tax allows for a deduction of 100% for machinery and equipment for power generation from renewable sources[63].

---

60 Another goal facing the program is to electrify 2,500 rural communities by the year 2012, would thus be mainly solar and wind energies.
61 It is the National Institute of housing of Mexican workers. The Infonavit's law which States that the contributions give right to obtain a credit for housing or the periodic return of the Fund which constitutes, so-called savings was passed in 1972. Institute of the National Housing Fund for workers.(n.d.) *Hipoteca Verde*. [Online] Available from: http://portal.infonavit.org.mx/wps/portal/TRABAJADORES [Accessed April 2012].
62 Secretaría de Energía. (n.d.) *Portal de Energías Renovables*. [Online] Available from: http://www.renovables.gob.mx/ [Accessed April 2012].
63 Art. 40: "...In this fraction shall apply provided that machinery and equipment are in operation or operation for a minimum period of 5 immediate years following the year in which the deduction is made except in the cases referred to in article 43 of this law. Taxpayers who fail to meet with the

13.2 CURRENT OVERVIEW OF RENEWABLE ENERGY IN MEXICO: PROGRESS ON THE INTRODUCTION OF WIND AND SOLAR ENERGY IN THE MEXICAN MARKET

13.2.1 Wind energy

Although widely considered to be a recent in innovation in Mexico wind energy can trace its origin in Mexico to 1988, when the government of President Carlos Salinas de Gortari granted the first permit to a wind farm in the Isthmus of Tehuantepec.

According to data from the Global Wind Energy Council, the total installed capacity in Mexico increased from 3 MW in 2005 to 519 MW in 2010. In 2009 and 2010 alone the increase was 156%, as shown in Table 13.3.

The Institute of Electrical Research (IIE) in Mexico estimates wind energy potential in the country at 71,000 MW, where plant capacity factors are above 20%. For plant capacity factors greater than 30% the estimated potential is 11,000 MW[64]. The region of the country and one of the world's greatest potential for wind energy is the Isthmus of Tehuantepec, in State of Oaxaca. It currently houses the majority of wind projects in operation with a capacity of 518.63 MW as of 2011 (see Figure 13.2). In addition, wind projects under construction and under development (open season 2010–2014) in the state of Oaxaca have a planned capacity of 2,149.1 MW upon completion.[65]

The states of Tamaulipas, Veracruz, Hidalgo, Nuevo Leon, Chihuahua, Coahuila, Sonora, Jalisco, Chiapas and Yucatan Peninsula also show promise for future wind projects. In the State of Baja California, the area known as La Rumorosa has an estimated wind potential of 5,000 MW (one of the nation's best). In these states wind projects under development total 4,871 MW. Of these projects in development, those located in Baja California is destined for energy export to the U.S. Under the current laws and objectives it should be mentioned that Mexico's currently main alternative to private participation in the development of wind energy are: Independent Power Production (projects tendered by CFE), the Remote Self-Supply and export.

But while wind has seen more development and growth has in recent years in Mexico than other renewable energies, it still has a long way to go. Mexico is not among the top ten countries with the largest installed capacity of wind energy.

minimum period specified in this paragraph shall cover, where appropriate, the corresponding tax by the difference resulting between the amount deducted in accordance with this fraction and the amount which should deduct in each exercise in terms of this article or article 41 of this law, have not implemented the deduction of 100%. For these purposes, the taxpayer shall submit supplementary declarations for each of the corresponding exercises, later than the month following that in which is not complied with the deadline established in this fraction, and must cover the surcharges and the corresponding update, from the date in the...." Cámara de Diputados. LXI Legislatura. (n.d.) *Ley del Impuesto sobre la Renta*. [Online] Available from: http://www.diputados.gob.mx/LeyesBiblio/ref/lisr. htm [Accessed April 2012].

64 Secretaría de Energía. (2011) *Prospectiva de Energías Renovables 2011–2025*. [Online] Available from: http://www.renovables.gob.mx/portal/Default.aspx?id=2094&lang=1 [Accessed April 2012].

65 Asociación Mexicana de Energía Eólica, A.C. (2011) *Panorama General de la Energía Eólica en México 2011*. [Online] Available from: http://www.amdee.org/Recursos/Proyectos_en_Mexico [Accessed April 2012].

Table 13.3 Total installed wind capacity in Mexico (Global Wind Energy Council, n.d.).

Year	2005	2006	2007	2008	2009	2010
MW	3	85	85	85	202	519

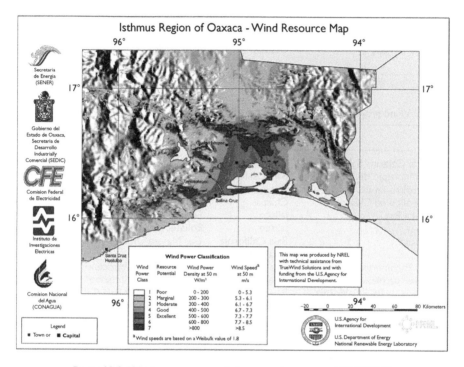

Figure 13.2 Wind resource map-isthmus region oaxaca MX AMDEE.

As of 2009 Mexico ranked in 24th place[66]. However, in regards to the manufacture of renewable energy products, there has been a significant increase in the development of domestic manufacturing of blades for wind turbines and industrial and power generator pylons, among others items[67]. The work of the Mexican Wind Energy Association (AMDEE) has been very important in promoting the development of the wind energy industry in the country, proposing solutions to address common problems of developers and both legal and public policy measures to encourage investment and growth of this industry.

66 Mexican Wind Energy Association, A.C. (2011) *Panorama General de la Energía Eólica en México 2011.* [Online] Available from: http://www.amdee.org/Recursos/Proyectos_en_Mexico [Accessed April 2012].

67 Ibid.

Table 13.4 Wind projects in operation – 2011 (Mexican Association on Wind Energy, 2011).

Project	Location	Developed by	MW
La Venta	Oaxaca	CFE	1.6
La Venta II	Oaxaca	CFE	83.3
Parques Ecológicos de México	Oaxaca	Iberdrola	79.9
Eurus, 1st Phase	Oaxaca	Cemex/Acciona	37.5
Eurus, 2nd Phase	Oaxaca	Cemex/Acciona	212.5
Gobierno de Baja California	Baja California	GBC/Turbo Power Services	10
Bii Nee Stipa I	Oaxaca	Cisa-Gamesa	26.35
La Mata- La Ventosa	Oaxaca	Eléctrica del Valle de México	67.5
		Total MW in Operation	518.63

Table 13.5 Wind projects under development (Mexican Association on Wind Energy, 2011).

Project	Location	Developed by	MW
Eólica Santa Catarina S.A. de C.V.	Nuevo León	Next Energy de México, S.A. de C.V.	22
Proyecto Municipio de Comondu	Baja California	Next Energy de México, S.A. de C.V.	16
Proyecto Eólico en BC	Baja California	Geomex, S.A. de C.V.	870
Proyecto Eólico en Chiapas	Chiapas	Geomex, S.A. de C.V.	39
Vaquerías-La Paz	Jalisco	Eoliatec de México	60
Chinanpas	Jalisco	Eoliatec de México	64
Unión Fenosa	Baja California	Gas Natura/Unión Fenosa	1000
Sempra	Baja California	Sempra	1000
Asociados Panamericanos	Baja California	Asociados Panamericanos	1000
Wind Power de México	Baja California	Wind Power de México	500
Fuerza Eólica de Baja California	Baja California	Fuerza Eólica de Baja California	300

Without a doubt, there remain two important barriers to be overcome if wind energy is to be advanced in the country. A study by the United States Agency for International Development can be summarized as follows:

1) The main barrier is the energy planning methodology used by the Federal Electricity Commission (CFE) scheme as the Independent Power Producer (IPP), which has led to minimal development of non-hydro renewables (Luengo & Oven, 2008).
2) Regarding self-supply generation in remote areas transmission networks are controlled by the CFE, which complicates the procedure to give network access to the licensees, in addition, the CFE sets charges for transmission service (Luengo & Oven, 2008).

The study recommends reviewing the experiences of the U.S. States of California and Texas[68] that have applied two regulatory mechanisms that are among the key

---

68 Only the State of Texas has more than one quarter of the total installed capacity of wind power in the U.S. with 10.1GW. Sawin, J. (2011) *Renewables, 2011 Global Status Report*. Renewable Energy Policy Network for the 21st Century Secretariat.

Table 13.6 Installed solar photovoltaic systems in Mexico
(National Association of Solar Energy, 2010).

Year	Additional installed capacity (MW)
Before 2001	13.209
2001	1.052
2002	0.1855
2003	0.6234
2004	0.9923
2005	0.5151
2006	1.0561
2007	0.901
2008	0.87241
2009	5.712
2010	3.502
Total Installed in 2010	28.62

success factors of wind energy in the U.S. These led the U.S. to be leader in total wind power generated in 2008.[69] These mechanisms include temporary subsidies of wind power through a tax credit for renewable energy production, and setting minimum goals for generation by renewable energy.

The study concludes that Mexico is in a unique position to advance the development of wind energy by the recent approval of the LAERFTE, but suggests that the Mexican government should establish mandatory targets for renewable generation to the CFE and Luz y Fuerza del Centro, to ensure adequate and sustainable financing mechanisms are available and promote the development of transmission lines to transmit electricity from wind energy, among other measures (Luengo & Oven, 2008).

13.2.2 Solar energy

The origins of solar energy in Mexico date back to 1977, when the first installation of photovoltaic cells and modules for rural telephones in the Sierra de Puebla was completed[70]. In 1980, the National Solar Energy Association (ANES) was created, which was chaired by leading members of the Mexican scientific community concerned with the goal of dissemination and use of renewable energy. The ANES is an outstanding reference for research and development of solar energy in the country.

From 1977 to the present, the implementation of solar energy has been a slow and gradual process. Until 2001, photovoltaic systems in Mexico only had a total

---

69 In the year 2008 the total installed capacity was 25,170 MW and USA led worldwide, but in 2010 the growth of the total installed capacity in this country was slower, peaking at 40.2 GW and placing it behind China that it leads the world market with 44.7 GW. Sawin, J. (2011) *Renewables, 2011 Global Status Report*. Renewable Energy Policy Network for the 21st Century Secretariat.
70 Asociación Nacional de Energía Solar. [Online] Available from: http://www.anes.org/anes/ [Accessed April 2012].

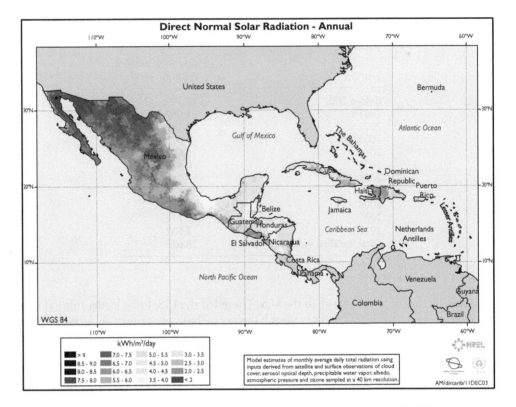

Figure 13.3 Direct normal solar radiation annual Mexico (Courtesy of NREL).

installed capacity of 14.26 MW. By 2010, the total installed countrywide capacity had doubled to 28.62 MW (see Table 13.6). The main uses have been for rural electrification, residential, water pumping, commercial, industrial, and isolated systems of transmission networks.

According to the Ministry of Energy, the average global solar irradiation of the country is 5 kWh/day/m² and can reach values between 8.0–8.5 kWh/day/m². Mexico's solar generation potential is the highest in the world where the Mexican government estimates an average near term installation potential, in the north and northwest regions of the country, of 1,653 MW. The government also calculates an estimated potential for solar hot water heating of more than 2,000,000 square meters annually.[71]

This should be taken advantage of by the Mexican authorities in order to reduce their dependence on fossil fuels and ensure energy security, as their actions to date have been insufficient in the support of the wide scale implementation of solar energy.

Only recently was the first solar combined cycle plant with a 14 MW solar field completed in the town of Agua Prieta, Sonora with a grant from the Global

71 Secretaría de Energía. (2011) *Prospectiva de Energías Renovables 2011–2025*. [Online] Available from: http://www.renovables.gob.mx/portal/Default.aspx?id=2094&lang=1 [Accessed April 2012].

Environment (GFE) under the World Bank.[72] Thanks to the Solar Heating Program (PROCALSOL) that has driven the current federal administration of 2010, 1,665,502 square meters of solar water heaters have been installed primarily for water heating for swimming pools, hotels, sports clubs, homes, hospitals and agricultural and industrial sectors.[73]

13.3   CONCLUSIONS

Mexico is heavily dependent on fossil fuels as primary sources to generate electricity (currently 60%). The Mexican government estimates an average annual addition of 2,399 MW by 2025 will be required to meet the power demand of the electric power utility. It is clear that the country needs to diversify its primary sources of energy production.

Mexico is blessed with great natural potential for geothermal, hydro, wind and solar power harvesting. However, it has not been until very recently that its energy policy has focused on promoting the development of domestic renewable energy. The creation of the LAERFTE was a key step in establishing a legal and regulatory framework for the use of renewable energy. The new law sets maximum participation goals for fossil fuel power generation for the years 2024 of 65%, 2035 at 60% and 2050 at 50%.

As of August of 2010, Mexico has a total installed capacity of 2,365 MW of renewable energy. After hydropower, wind power has seen the most growth in recent years, although this is still limited at 519 MW of wind capacity installed as of 2010.

Mexico is considered an emerging market for renewable energy, with great potential for many types of renewable energy and opportunities for clean energy exports to other countries (Howse & Bork, 2006). However, it faces important barriers to be overcome if advancement in the development of renewable energy for global market leadership is to be realized.

The Mexican Association of Wind Energy and Solar have provided solutions to straighten and define the path. However, it is important that the Mexican authorities on energy take an in depth look at the experiences of the world's leading countries in the market for renewable energies such as wind and solar, and learn from both the failures and successes.

A substantial part of the success of the United States has been the use of fiscal instruments such as regulatory mechanisms, among them tax credits (eg Renewable Energy Production Incentives, Clean Renewable Energy Bonds, Renewable Electricity Production Tax Credits, Modified Accelerated Cost-Recovery Systems, and Residential Solar, among others) and subsidies (Howse & Bork, 2006). However strong government support for the research of alternative energy sources since the 1970s through groups such as the National Renewable Energy Laboratory (NREL) has been key.

72 Comisión Federal de Electricidad. (n.d.) *Informe Anual 2010*. [Online] Available from: http://www.cfe. gob.mx/QuienesSomos/publicaciones/Paginas/Publicaciones.aspx [Accessed April 2012].
73 Asociación Nacional de Energía Solar. (2010) *Balance de Energía*. [Online] Available from: http:// www.anes.org/anes/index.php?option=com_wrapper&Itemid=13 [Accessed April 2012].

Mexico needs to achieve energy diversification and security. This will be a key task of the next federal government set to begin on December 1, 2012. Whatever the governing party, it is not only desirable that the next government have a continuity of programs and measures, but that it considers clean energy a great opportunity to create jobs, attract investment, export energy, ensure energy independence of the country. Most importantly, the next government should be mindful of protecting the environment through renewable energy development.

Chapter 14

Market study: Current state of the US solar market

Brett Prior

In 2011, there were over 1,800 megawatts (MW) of solar installed in the United States. Approximately 40% of the total volume, or 700 MW, was in the utility segment. This chapter of the book will attempt to answer the following key questions:

– What is the current state of the US utility solar market?
– What does the future hold for the US utility segment?

14.1 SOLAR PROJECTS IN OPERATION

14.1.1 Solar projects by technology

14.1.1.1 Crystalline silicon and thin film photovoltaics

There are over 1,000 MW of operational utility photovoltaic (PV) plants in operation in the US. The five largest plants are detailed in Table 14.1. The largest PV projects currently in operation will soon be dwarfed by the projects under construction and nearing completion, as many of these are over 300 MW. Four of the five largest projects in operation in the US are in the US Southwest and West, due to superior solar resource and more aggressive Renewable Portfolio Standards (RPS). The one exception is the Long Island Solar Farm, which is located in an area with high electricity rates.

14.1.1.2 Concentrating solar power

There are over 500MW of operational concentrating solar power (CSP) plants in the US. A list of the largest operational plants is stated in Table 14.2. The Solar Energy Generating Systems (SEGS) plants, phases I to IX, are parabolic trough plants developed by Luz, and they represent the largest solar complex in the world. They were built in the 1980s and later sold to NextEra and Cogentrix. The Martin Next Generation Solar Energy Center (NGSEC) plant was developed, built, and owned by the local utility, FP&L in Florida. It is co-located next to the largest fossil fuel thermal plant in the US (burning gas and oil), so they are able to feed the steam into the existing steam turbines. The Nevada Solar One plant is another trough project, originally developed by Solargenix, and now owned by Acciona.

Table 14.1 Sample of the largest crystalline silicon and thin film PV projects in the U.S. (Operating).

MW	Name	Location	Developer	Owner
45	Avenal	CA	NRG, Eurus	NRG, Eurus
40	Copper Mountain	NV	Sempra Generation	Sempra Generation
37.5	Mesquite Solar I	AZ	Sempra Generation	Sempra Generation
32	Long Island Solar Farm	NY	BP Solar	BP Solar
30	San Luis Valley Solar	CO	Iberdrola	Iberdrola
30	Cimarron I Solar	NM	First Solar	Southern Company, Turner Renewable Energy

14.1.1.3   Concentrating photovoltaic

There are only a handful of utility concentrating photovoltaic (CPV) plants in operation representing less than 40 MW with the largest, Amonix's 30 MW power plant in Colorado, making up the vast majority.

14.1.2   Solar projects in operation by state

For the ~1,000 MW of operational utility PV plants in the US, nearly 40% are located in California, with Arizona, Nevada, Florida, and New Mexico representing the majority of the rest. The SEGS projects help California to reach nearly 40% of all utility projects. Arizona benefits from the 37.5 MW Mesquite Solar I project. Nevada has the Nevada Solar One (64 MW) project and the 40MW Copper Mountain. Florida has the Martin CSP project, as well as some large PV projects, namely DeSoto (25 MW) and Space Coast (10 MW). New Mexico has the 30 MW Cimarron First Solar project amongst others.

14.1.3   Leading solar developer for completed solar projects

For the utility projects that have been built, the developer with the largest share is a developer who went bankrupt in 1991, Luz International. But, their legacy lives on in the 354 MW SEGS trough plants which continues to hold the title as the world's largest solar project.

The second largest developer of completed utility solar plants is SunEdison, the developer founded by Jigar Shah and Claire Broido Johnson in 2003. The largest of SunEdison's completed plants is the 30 MW Webberville project in Texas. It should be noted that SunEdison is now owned by MEMC, a manufacturer of polysilicon and silicon wafers used in the production of PV cells.

First Solar places third in terms of operational US utility solar projects with 121 MW. Their two largest completed projects are the 30 MW Cimarron project in New Mexico, and the 22 MW Blythe project in California. First Solar was one of the first module manufacturers to expand into the project development (systems) business, and has been aggressively pushing forward as an acquirer of project pipelines as well (they acquired OptiSolar pipeline in 2009, and NextLight in 2010).

NextEra Energy Resources (NER) is a subsidiary of NextEra Energy which is also the parent company of Florida Power & Light, the largest utility in Florida. NER is

Table 14.2 Sample of the largest CSP projects in the U.S. (Operating).

MW	Name	Location	Developer	Owner
354	SEGS	CA	Luz	NextEra/ Cogentrix
75	Martin NGSEC	FL	FP&L	FP&L
64	Nevada Solar One	NV	Solargenix	Acciona

Table 14.3 Percent of total US utility market: Southwest states compared to rest of U.S.

State	% of total US utility market
California	39%
Arizona	12%
Nevada	10%
Florida	9%
New Mexico	8%
Rest of the US	22%

North America's largest owner and operator of wind and solar projects. Considering just the operational utility solar plants, NER ranks 4th with about 115 MW of capacity. This includes the Martin NGSEC at 75 MW, DeSoto PV at 25 MW, and Space Coast PV at 10 MW.

Sempra Generation is a subsidiary of Sempra Energy which is also the parent company of SDG&E, one of the largest California utilities. The Sempra Generation division owns and operates power plants including wind, solar, and natural gas plants. Sempra Generation ranks as the 5th largest developer of operational US utility solar plants, as it has developed both the 40 MW Copper Mountain as well as the 38 MW Mesquite Solar I projects.

14.1.4 Leading debt providers for completed solar projects

Provider	Debt provided
Rabobank	$50M
Union Bank	$38M
Credit Agricole	$35M
Mizuho	$35M
Natixis	$35M
Santander	$35M
Sumitomo Mitsui	$35M
UniCredit	$35M

For the completed solar projects in the US, the debt funding has often come from major European and Japanese banks. For example, the 45 MW Avenal project sourced

Table 14.4 Developers and sizes of completed US utility solar market by percent.

Developer	MW of completed projects	% of total completed US utility solar market
Luz	354	22%
SunEdison	141	9%
First Solar	121	8%
NextEra Energy	115	7%
Sempra Generation	96	6%
All other developers	779	48%

its debt from a consortium of 6 banks including: Natixis, UniCredit, Credit Agricole, Mizuho, Santander, and Sumitomo Mitsui. An example of PV project debt financed by a US bank is the 10 MW LS Power White Oak project, which received its debt funding from Union Bank (although Union Bank is a subsidiary of Mitsubishi UFJ).

14.1.5   Leading tax equity providers for completed solar projects

Provider	Tax equity provided
Wells Fargo	$100M
Bank of America	$87M
Metlife	$87M
JP Morgan	$60M
US Bank	$52M
East West Bank	$28M

The largest contributors to tax equity were the companies that created funds for projects, such as Wells Fargo's $100 million fund for GCL Solar. Other notable players were investment bank Bank of America and insurance giant Metlife which provided lease financing for SunEdison's 15 MW Davidson project.

14.1.6   Leading direct equity providers for completed solar projects

Provider	Direct equity provided
NRG Energy	$37M
Eurus Energy	$16M

The leading providers of direct equity for completed utility-scale solar projects have been Independent Power Producers (IPPs) such as NRG, developers such as Eurus and Acciona, and utilities.

14.2   BOTTOM-UP VIEW OF THE US UTILITY-SCALE SOLAR MARKET, 2012–2015

14.2.1   Solar projects under development or construction

In addition to the ~2 GW of operational utility solar projects in the US, there are another 13 GW of utility solar projects with signed power purchase agreements (PPAs) that are still under development or construction, and expected to be completed between 2012 and 2015. While many of these projects will be completed on schedule, many of these projects will never reach the finish line. Potential obstacles that may prevent a utility project with a signed PPA from being completed include:

–   Inability to secure financing (due to unattractive economics, technology risks, or other investor concerns)
–   Inability to secure required permits
–   Inability to secure interconnection
–   Inability to receive Public Utility Commission (PUC) approval; perhaps due to uncompetitive PPA pricing

While it is difficult to predict the percentage of utility solar projects that will be completed, the large number of projects that have already secured financing and begun construction gives some certainty that the market will continue to grow substantially, even if the success rate for development stage projects is low.

14.2.2   Crystalline silicon and thin film PV

Some of the largest PV projects are being developed by two manufacturers with downstream arms: First Solar and SunPower. Both have arranged loans guarantees from the Department of Energy to provide debt for these megaprojects, and they are now in the process of locking down tax equity (Topaz and Desert Sunlight already have new equity owners) and constructing several of these projects.

14.2.3   CSP

The largest CSP projects are being developed by two system manufacturers with in-house development teams: BrightSource and Abengoa. Ivanpah, Solana, and Mojave Solar have all secured loans from the Department of Energy (DoE), and have secured equity financing and begun construction. Financing for some of the other mega-CSP projects is still outstanding and may prove more challenging without the subsidized debt from the DoE.

14.2.4   CPV

Soitec, Amonix, and SolFocus are developing several mega-CPV projects in the US Southwest. In particular, Soitec has signed PPAs for 305 MW of projects with

Table 14.5 Large crystalline silicon and thin film PV projects under development.

MW-ac	Name	Location	Developer	Owner
550	Topaz	CA	First Solar	MidAmerican Energy Holdings
325	Rosamond County	CA	SunPower	SunPower
325	Antelope Valley	CA	SunPower	SunPower
300	Desert Sunlight	CA	First Solar	NextEra Energy, GE EFS
300	Desert Stateline	CA	First Solar	

SDG&E and is awaiting PUC approval and financing before commencing construction. Amonix has a 30 MW project in Colorado slotted for completion in May 2012, and SolFocus is providing systems for 30 MW worth of PPAs in SDG&E.

14.2.5   Solar projects by state

California is likely to remain the center of activity for utility scale PV project development, due its aggressive 33% RPS requirement by 2020, and it massive electricity consumption. Arizona and Nevada are distant second and thirds, with some of the projects sited in Nevada potentially slotted to serve California utilities.

14.2.6   Leading solar developer

As mentioned in earlier sections, the leaders CSP developers are BrightSource and Abengoa. On the PV side, project leaders include 2 vertically-integrated manufacturers (First Solar and SunPower), and two IPPs (Sempra Generation and NextEra).

14.2.7   Leading debt providers

Debt providers	Solar 2011 Investment
Federal Financing Bank	$6,815M
Bank of America	$1,750M
Citigroup	$730M
Goldman Sachs	$730M
BBVA	$426M
Credit Suisse	$426M
North American Development Bank	$77M
Macquarie Energy	$41M
Seminole Financial	$4M

Debt has already been committed to many of the largest solar projects – with expected commercial operation dates ranging from 2012 to 2016. By far the largest debt provider is the Federal Financing Bank, which is entity that provides the loans through the 1705 Department of Energy Loan Guarantee program. Bank of America ranks

Table 14.6 Large CSP projects under development.

MW	Name	Location	Developer	Owner
1,200	SCE XVI-XXI	CA	BrightSource	BrightSource
370	Ivanpah	CA	BrightSource	BrightSource, Google, NRG
280	Solana	AZ	Abengoa	Abengoa
270	Hidden Hills I	CA	BrightSource	BrightSource
270	Hidden Hills II	CA	BrightSource	BrightSource
250	Abengoa Mojave Solar	CA	Abengoa	Abengoa

Table 14.7 Solar projects by state.

State	GW of solar with PPAs	% of total US utility market projects with PPAs
CA	9.3	75%
AZ	1.2	10%
NV	1.0	8%
HI	0.3	3%
FL	0.1	1%
Rest of the US	0.5	4%

Table 14.8 Leading solar developers.

Developer	MW of projects with PPAs under development/construction	% of all US utility solar projects with PPAs under development/ construction
BrightSource	2,600	20%
First Solar	2,185	17%
SunPower	1,008	8%
Sempra Generation	549	4%
Abengoa	530	4%
NextEra Energy	508	4%
All other developers	6,150	43%

second, thanks largely to its $1.4 billion commitment to NRG/Prologis's 700 MW Project Amp. Citigroup and Goldman Sachs rank well due to their commitment to provide debt (guaranteed by the DoE's 1705 FIPP program) for First Solar's 550 MW Desert Sunlight project. Similarly, BBVA and Credit Suisse make the top 10 thanks to debt provided to another loan guarantee recipient, NextEra's 250 MW Genesis CSP project.

14.2.8   Leading tax equity providers

Tax equity providers	Solar 2011 Tax Equity investment
Exelon	$714M
GE EFS	$574M
MidAmerican	$408M
Wells Fargo	$200M
Google	$168M

The primary tax equity providers for the projects due to come online in the next several years typically fall into one of four buckets: investment banks, insurance companies, utilities, and new entrants. Exelon, a utility covering Illinois, Pennsylvania, and Maryland, has entered this space in a big way with its equity investment into First Solar's 230 MW Antelope Valley. Similarly, General Electric's Energy Financial Services (EFS) has indicated its potential appetite for tax equity by taking a stake in First Solar's 550 MW Desert Sunlight. Also noteworthy is Warren Buffet's MidAmerican taking an equity stake in First Solar's 290 MW Agua Caliente project.

14.2.9   Leading direct equity providers

Direct equity providers	Solar 2011 direct equity provided
NRG	$2,381M
NextEra	$1,242M
Abengoa	$400M
Sempra	$265M
SolarReserve	$246M
Cogentrix	$54M

Direct equity for utility solar plants will likely continue to come from two main sources: IPPs and developers. By far the largest equity investor in future utility solar projects is NRG Solar, with commitments to BrightSource's Ivanpah, SunPower's CVSR, Prologis's Project Amp, and First Solar's Agua Caliente. The second largest provider of equity to future solar projects is NextEra Energy Resources with commitments to First Solar's Desert Sunlight, and their own Genesis project.

14.3   TOP-DOWN VIEW OF THE US UTILITY-SCALE SOLAR MARKET, 2012–2015

An alternate way of looking at the future of the US utility solar market is top down – that is, based on macro drivers of demand rather than aggregating individual company plans.

14.3.1 Total US electricity demand and supply growth

In general terms, US electricity demand is not currently a growing market. Compared to historical energy growth over the past few decades, the demand level has changed due to the dual impacts of energy efficiency and the shrinking size of the manufacturing sector meaning that electricity consumption per capita in the US will continue to decline. That said, the current fleet of coal and nuclear plants that power the US are aging and will need to be replaced over the coming decades. In recent years, wind and natural gas plants have each represented close to 40% of all new electricity capacity additions. With the best wind sites developed, and the production tax credit (PTC) slated to expire at the end of 2012, it is likely that wind's share of new capacity will drop. It remains an open question whether solar will be able to step up a take a larger share of the new electricity capacity added in the US in future years, or whether natural gas will fill the void left by retiring coal plants.

14.3.2 RPS requirements by state

For the next decade, utility-scale solar will continue to be driven by state-level RPS mandates. California's 33% requirement by 2020 is a key example of a state policy that is essential to the growth of the segment. Other key states for utility-scale solar, Arizona and Nevada, also have RPS policies in place, although not as ambitious as California's target. Arizona's RPS requires 15% renewables by 2025, while Nevada's RPS requires 25% by 2025. In every state there remains a potential risk and potential upside that the legislature may either decrease or increase the RPS requirement. This happened in California in 2011, when the 2020 target was increased to 33%.

14.4 KEY TRENDS IN THE US UTILITY-SCALE SOLAR MARKET

14.4.1 End of the section 1603 cash grant program

In order to meet RPS requirements, there has been a rush of utility scale solar project development. Many of these projects have secured the necessary permits, interconnection agreements, and even signed PPAs from utilities. The final step is financing. In 2011, two program deadlines caused a mad dash as many of these shovel-ready projects sought to secure either a Department of Energy loan guarantee before September 30, or they sought to qualify for the section 1603 cash grant program (with a deadline of December 31). With both of these deadlines now past, the financing landscape has returned to "normal." While this isn't necessary a shortage of debt or tax equity for solar projects, there are definitely more projects seeking financing than the market can support. The end result is that debt and tax equity providers can be selective, and choose only the deals with the lowest risk profiles, and highest returns. This means that deals with "hair on them" – that is, any project with issues may find itself at the back of the evaluation queue. Issues that can derail the financing process include: technology risk, less credit-worthy off-takers, a low PPA price, uncertain O&M expenses, and anything else that might put the equity investor's internal rate of return (IRR) at risk.

14.4.2   IPPs moving from natural gas and wind to solar

Over the last decade, several of the largest independent power producers (IPPs) moved from coal and natural gas plants to wind farms, as utilities demanded wind projects to meet state RPS requirements. These same IPPs are now moving into utility scale solar in a big way. Four major examples are NRG Energy, NextEra Energy Resources, AES, and enXco. Each of these companies is well positioned to play a leading role in solar project development. IPPs are logical owners of utility solar projects for several reasons. First, they have strong relationships with the utilities. Second, as large firms with experience bringing complex projects online they are able to raise debt from large banks to finance the solar projects. Finally, IPPs have large annual profits that allow them to effectively utilize the tax credits and depreciation that accrues to the owner of large solar projects.

14.4.3   Once the RPS requirements are met, then what? or – can solar compete with natural gas plants?

The system price for utility solar has improved significantly over the past half-decade and should continue to improve going forward. While this puts solar power on a trajectory to close the gap in the cost differential between solar and natural gas, it does not address one key shortcoming of solar power: intermittency. Gas plants can be turned on and off as needed and can be depended upon to generate power at times of peak demand. The same cannot be said for solar power plants.

The addition of energy storage to a solar plant would give it dispatch-ability, but the additional cost of the storage would make the plant unpalatable to cost-conscious utilities and PUCs. While there are technologies in the lab that offer the hope of much lower cost energy storage, it would be many years before these technologies could be commercialized and incorporated into utility scale solar plants. Accordingly, solar will remain at a severe disadvantage when trying to compete head-to-head with natural gas plants. It is likely that once the RPS requirements are met, utilities will only procure additional solar if the case can be made that solar is an economically superior option to the alternatives. With natural gas prices at record low levels, solar would be hard-pressed to compete on a strictly levelized cost of electricity basis in the United States. Only time will tell to see how solar energy is going to be able to co-exist with conventional sources of energy.

References

APS (2010). *2010 Request for Proposal ("RFP") for Renewable Energy Small Generator Resources*. [Online] Available from: http://www.aps.com/files/_files/rfp/2010SmallGen_RFP.pdf [Accessed 21st May 2012].

Arndt, M. (1989) *Gasoline Price Climbs Faster Than in '70s*. [Online] Available from: http://articles.chicagotribune.com/1989-04-11/news/8904030004_1_gasoline-prices-oil-industry-analyst-william-randol [Accessed 24th May 2012].

Businessweek (2012). "Buffett Plans More Solar Bonds After Oversubscribed Deal", *http://www.businessweek.com/news/2012-02-29/buffett-plans-more-solar-bonds-after-topaz-deal*

California Energy Commission. (n.d.) *The Energy Almanac*. [Online] Available from: http://energyalmanac.ca.gov/electricity/us_per_capita_electricity-2010.html [Accessed 20th May 2012].

California Legislature (2011), *Senate Bill 2*. [Online] Available from: www.leginfo.ca.gov/pub/11-12/bill/sen/sb_0001-0050/sbx1_2_bill_20110412_chaptered.pdf [Accessed 9th April 2012].

California Valley Solar Ranch (2009) California Valley Solar Ranch Fact Sheet. [Online] Available from: http://www.californiavalleysolarranch.com/Fact_Sheet.pdf [Accessed 21st May 2012].

Cámara de Diputados. LXI Legislatura. (n.d. – a) *Ley de la Comisión Reguladora de Energía*. [Online] Available from: http://www.diputados.gob.mx/LeyesBiblio/ref/lcre.htm [Accessed 8th March 2012].

Cámara de Diputados. LXI Legislatura. (n.d. – b) *Ley del Impuesto sobre la Renta*. [Online] Available from: http://www.diputados.gob.mx/LeyesBiblio/ref/lisr.htm [Accessed 8th March 2012].

Cámara de Diputados. LXI Legislatura. (n.d. – c) *Ley para el Aprovechamiento de Energías Renovables y el Financiamiento de la Transición Energética*. [Online] Available from: http://www.diputados.gob.mx/LeyesBiblio/ref/laerfte.htm [Accessed 8th March 2012].

Collins, F. (2010) *Utility Scale PV Plant Optimization by Senior Systems Engineer Forrest Collins of juwi solar Inc.* [presentation] Utility Scale PV Yield Optimization Conference and Expo US, November 30.

Connolly, K. (2012) *Germany to cut solar power subsidies*. [Online] Available from: www.guardian.co.uk/world/2012/mar/02/germany-cuts-solar-power-subsidies [Accessed 9th April 2011].

Couture, T. (2011), GuestPost: Spain's Renewable Energy Odyssey. [Online] Available from: http://www.greentechmedia.com/articles/read/spains-renewable-energy-odyssey/ [Accessed 21st May 2012].

DBCCA (2011). "*Get Fit Plus: Derisking Clean Energy Business Models in a Developing Country Context.*" http://europa.eu/epc/pdf/workshop/background_get_fit_plus_final_040711_en.pdf

Department Of Conservation, Division of Land Resource Protection (2011) *Solar Power and the Williamson Act.*

Douglass, E. (2006) Calpine Bankruptcy Filing Came as It Amassed $1 – Billion '05 Loss. [Online] Available from: http://articles.latimes.com/2006/may/20/business/fi-calpine20 [Accessed 13th May 2012].

EIA (2011) International Energy Statistics. [Online] Available from: http://www.eia.gov/countries/data.cfm EIA [Accessed 13th May 2012].

Feldman, D., Mendelsohn, M. & Coughlin, J. (2012). "The Technical Qualifications for Treating Photovoltaic Assets as Real Property by Real Estate Investment Trusts (REITs)" NREL, 8 pages

Global Finance (2010), *Global Finance names the World's 50 Safest Banks 2010*, September 2010.

Global Wind Energy Council. (n.d.) *Regions. Mexico.* [Online] Available from: http://www.gwec.net/index.php?id=19 [Accessed 14 March 2012].

Harper, J., Karcher, M. & Bolinger, M. (2007), *Wind Project Financing Structures: A Review and Comparative Analysis*, Lawrence Berkeley National Laboratory. eetd.lbl.gov/ea/emp/reports/63434.pdf

Hicks, D. (2010) *NV Energy 2010 Request for Proposals for Renewable Resources: RFP Protocol and Bid Information.* [Online] Available from: https://www.nvenergy.com/company/doingbusiness/rfps/images/NVE2010_RFP_Bidders_Final_3-11-10.pdf [Accessed 21st May 2012].

Howse, R. & Bork, P. (2006) *Opportunities and Barriers for Renewable Energy in NAFTA.* [Lecture] Third North American Symposium on Assessing the Environmental Effects of Trade, Commission for Environmental Cooperation, Montreal, February 2006.

Luengo, M. & Oven, M. (2008) *Análisis comparativo del marco eléctrico legal y regulatorio de EEUU y México para la promoción de la energía eólica. U.S. International Development Agency.* [Online] Available from: *http://www.amdee.org/Publicaciones/publicaciones* [Accessed 5th March 2012].

Mendelsohn, M. & Harper, J. (2012). *§1603 Treasury Grant Expiration: Industry Insight on Financing and Market Implications*, NREL, June. [Online] Available from: http://www.nrel.gov/docs/fy12osti/53720.pdf

Mendelsohn, M. & Hubbell, R. (2012). *NREL's Renewable Energy Finance Tracking Initiative.* Presented via webinar, April 26, 2012. [Online] Available from: https://financere.nrel.gov/finance/REFTI

Mendelsohn, M. & Kreycik, C. (2012). *Federal and State Structures to Support Financing Utility-Scale Solar Projects and the Business Models Designed to Utilize Them.* National Renewable Energy Laboratory Technical Report TP-6A2-48685. April. [Online] Available from: http://www.nrel.gov/docs/fy12osti/48685.pdf

Mendelsohn, M., Kreycik, C., Bird, L., Schwabe, P. & Cory, K. (2012a). *Utility-Scale Concentrating Solar Power and Photovoltaic Projects: A Technology and Market Overview.* National Renewable Energy Laboratory Technical Report TP-6A20-53086. March. [Online] Available from: http://www.nrel.gov/docs/fy12osti/53086.pdf

Mendelsohn, M., Lowder, T. & Canavan, B. (2012b); *Utility-Scale Concentrating Solar Power and Photovoltaic Projects: A Technology and Market Overview.* National Renewable Energy Laboratory Technical Report TP-6A20-51137. April. [Online] Available from: http://www.nrel.gov/docs/fy12osti/51137.pdf

Mendelsohn, M. (2011a), Analysis completed for presentation to DOE – not published.

Mendelsohn, M. (2011b), "Renewable Generation Project Deployment: Policy and Business Model Interaction", Presented to [ ] conference, Shanghai, China, November 7, 2011.

Mendelsohn, M. (2011c) *Looking Under the Hood: Some Perspective on the Loan Guarantee Program*, NREL, December 2011, [Online] Available from: https://financere.nrel.gov/finance/content/looking-under-hood-some-perspective-loan-guarantee-program

Mendelsohn, M. (2012a) *Best of Both Worlds: What if German installation costs were combined with the best solar resources?* NREL, April 2012, [Online] Available from: https://financere.nrel.gov/finance/content/germany-solar-feed-in-tariff-FIT-insolation-resource-comparison

Mendelsohn, M. (2012b) *Will Solar Projects Need Tax Equity in the Future? Yes, but Baby Steps Toward Securitization Improve the Situation.* NREL Renewable Energy Project Finance website article. April. [Online] Available from: https://financere.nrel.gov/finance/content/solar-PV-photovoltaics-value-tax-credits-equity-accelerated-depreciation-securitization

Mendelsohn, M. (2012c) *Tapping the Capital Markets: Are REITs Another Tool in Our Toolbox?* National Renewable Energy Laboratory Renewable Energy Project Finance website article. February. [Online] Available from: https://financere.nrel.gov/finance/content/capital-markets-reit-real-estate-investment-trust-renewable-energy-project-finance-prologis-KIMCO

Mexican Association on Wind Energy. (2011) *Panorama de la Energía Eólica en México 2011.* [Online] Available from: http://www.amdee.org/Recursos/Proyectos_en_Mexico [Accessed 5th March 2012].

Mintz, L., Cohn, F. & Glovsky, P. PC, produced in collaboration with GTM Research *Renewable Energy Project Finance in the U.S.: 2010–2013 Overview and Future Outlook,* January 2012. [Online] Available from: http://www.mintz.com/publications/3055/Renewable_Energy_Project_Finance_in_the_US_20102013_Overview_and_Future_Outlook

Monast, J., Anda, J. & Profeta, T. (2009) *U.S. Carbon Market Design: Regulating Emission Allowances as Financial Instruments.* [Online] Available from: http://www.nicholas.duke.edu/ccpp/ccpp_pdfs/carbon_market_primer.pdf [Accessed 22nd May 2012].

National Association of Solar Energy. (2010) *Balance de Energía.* [Online] Available from:http://www.anes.org/anes/index.php?option=com_wrapper&Itemid=13 [Accessed 5th March 2012].

National Institute of Housing of Mexican Workers. (n.d.) *Hipoteca Verde.* [Online] Available from: http://portal.infonavit.org.mx/wps/portal/TRABAJADORES [Accessed 14th March 2012].

Nickerson, C., Ebel, R., Borchers, A., & Ng Carriazo, F. (2007) *Major Uses of Land in the United States.* [Online] Available from: www.ers.usda.gov/publications/eib89 [Accessed 1st May 2012].

Ng, P. (2009) *Draft Unit Cost Guide for Transmission Lines.* [Presentation] Folsom, 26th February.

PG&E (2012) *Wholesale Generator Interconnections.* [Online] Available from: http://www.pge.com/b2b/newgenerator/wholesalegeneratorinterconnection/index.shtml [Accessed 23rd May 2012].

Population Division of the Department of Economic and Social Affairs of the United Nations Secretariat (2009) World Population Prospects: The 2008 Revision. [Online] Available from: http://www.un.org/esa/population/publications/popnews/Newsltr_87.pdf [Accessed 13th May 2012].

Prior, B. (2010) Update: Ivanpah Gets Approval; Construction May Begin Soon. [Online] Available from: http://www.greentechmedia.com/articles/read/brightsource-eagerly-awaits-final-decision-from-the-california-energy-commi/ [Accessed 21st May 2012].

Reuters (2012) *Solarhybrid says to file for insolvency.* [Online] Available from: www.reuters.com/article/2012/03/20/solarhybrid-brief-idUSL6E8EKALM20120320 [Accessed 9th April 2011].

Sandoval, J.C., Bosl, B. & Eckermann, A. (eds.) (2006) *Energías Renovables para el Desarrollo Sustentable en México.* Mexico, Secretaría de Energía. [Online] Available from: http://www.sener.gob.mx/portal/publicaciones.html [Accessed 6th March 2012].

Sawin, J. (2011) *Renewables, 2011 Global Status Report.* Renewable Energy Policy Network for the 21st Century Secretariat.

Schwabe, P., Mendelsohn, M., Mormann, F. & Arent, D. (forthcoming). "Mobilizing Public Capital for Renewable Energy Project Finance: Insights from Expert Stakeholders", NREL, Stanford University, JISEA, 13 pages.

The Secretary of Energy. (2011) Prospectiva de Energías Renovables 2011–2025. [Online] Available from: http://www.renovables.gob.mx/portal/Default.aspx?id=2094&lang=1 [Accessed 4th March 2012].

The Secretary of Energy. (n.d.) *Portal de Energías Renovables*. [Online] Available from: http://www.renovables.gob.mx/ [Accessed 4th March 2012].

SolarServer (2012), "German PV installations in 2011 even higher than in record year 2010; 3 GW installed in December", Accessed June 20, 2012. http://www.solarserver.com/solar-magazine/solar-news/current/2012/kw02/german-pv-installations-in-2011-even-higher-than-in-record-year-2010-3-gw-installed-in-december.html

Thompson Reuters (2012) *Cushing, OK WTI Spot Price FOB* [Online] Available from: http://www.eia.gov/dnav/pet/hist/LeafHandler.ashx?n=pet&s=rwtc&f=m [Accessed 11th August 2012].

Wesoff, E. (2011) Bechtel on BrightSource's Ivanpah CSP Build. [Online] Available from: http://www.greentechmedia.com/articles/read/Bechtel-on-BrightSources-Ivanpah-CSP-Build/ [Accessed 21st May 2012].

Wesoff, E. (2011) "DOE's Chu and SunPower's Swanson on the SunShot Initiative", February 2011, http://www.greentechmedia.com/articles/read/does-chu-and-sunpowers-swanson-on-the-27m-sunshine-initiative/

USDA (2012) [Online] *State Facts Sheet:* California. Available from: www.ers.usda.gov/State-Facts/CA.htm [Accessed 1st May 2012].

USEPA (2011) *National Environmental Policy Act Basic Information.* [Online] Available from: http://www.epa.gov/oecaerth/basics/nepa.html [Accessed 16th April 2012].

Index

Printed and bound by CPI Group (UK) Ltd, Croydon, CR0 4YY

24/10/2024

01778287-0006